The U.S.-Mexican Border Environment

Lining the All-American Canal: Competition or Cooperation for the Water in the U.S.-Mexican Border?

El Colegio de la Frontera Norte

CONSORCIO DE INVESTIGACION Y POLITICA AMBIENTAL DEL SUROESTE
SOUTHWEST CONSORTIUM FOR ENVIRONMENTAL RESEARCH & POLICY

SCERP Monograph Series, no. 13

A series edited by Paul Ganster

Contributors

José Luis Castro Ruíz	El Colegio de la Frontera Norte
Alfonso Andrés Cortez Lara	Michigan State University
Peter Culp	Sonoran Institute
María Rosa García Acevedo	California State University, Northridge
Gerardo García Saillé	Universidad Autónoma de Baja California
Jaime Herrera Barrientos	Centro de Investigación Científica y de Educación Superior de Ensenada
Osvel Hinojosa Huerta	Pronatura Sonora
Doris Jorquera Flores	El Colegio de la Frontera Norte-Centro de Investigación Científica y de Educación Superior de Ensenada
Ángel López López	Universidad Autónoma de Baja California
Donna Lybecker	Pennsylvania State University Altoona
Fernando A. Medina Robles	Universidad Autónoma de Baja California
Stephen P. Mumme	Colorado State University
J. A. Navarro Urbina	Comisión Nacional del Agua
M. Norzagaray Campos	Centro Interdisciplinario de Investigación para el Desarrolló Integral Regional-Sinaloa
Vicente Sánchez Munguía	El Colegio de la Frontera Norte
Francisco Raul Venegas Cardoso	Universidad Autónoma de Baja California
Francisco Zamora Arroyo	Sonoran Institute

The Southwest Consortium for Environmental Research and Policy (SCERP) is a consortium of U.S. and Mexican universities dedicated to addressing environmental issues of the U.S.-Mexican border region through applied research, outreach, and regional capacity building.

SCERP Universities
Arizona State University
El Colegio de la Frontera Norte
Instituto Tecnológico de Ciudad Juárez
Instituto Tecnológico y de Estudios Superiores de Monterrey
New Mexico State University
San Diego State University
Universidad Autónoma de Baja California
Universidad Autónoma de Ciudad Juárez
University of Texas at El Paso
University of Utah

SCERP website: www.scerp.org

The U.S.-Mexican Border Environment

Lining the All-American Canal: Competition or Cooperation for the Water in the U.S.-Mexican Border?

Edited by Vicente Sánchez Munguía

San Diego State University Press
Southwest Consortium for Environmental Research and Policy (SCERP)
El Colegio de la Frontera Norte (COLEF)

Published by
San Diego State University Press
5500 Campanile Drive
San Diego, CA 92182-8141
http://sdsupress.sdsu.edu

Cover photos by U.S. Bureau of Reclamation Lower Colorado Region

Printed in the United States of America

ISBN 0-925613-49-5

Previously published volumes in the SCERP Monograph Series,
The U.S.-Mexican Border Environment

No. 1 A Road Map to a Sustainable 2020
No. 2 Water Issues along the U.S.-Mexican Border
No. 3 Economy and Environment for a Sustainable Border Region
No. 4 U.S.-Mexican Border Communities in the NAFTA Era
No. 5 Overcoming Vulnerability: The Southwest Center for Environmental Research and Policy's Research Program (1990–2002) and Future Agenda
No. 6 Air Quality Issues along the U.S.-Mexican Border
No. 7 Trade, Energy, and the Environment: Challenges and Opportunities for the Border Region, Now and in 2020
No. 8 Binational Water Management Planning
No. 9 Tribal Environmental Issues of the Border Region
No. 10 Improving Transboundary Air Quality with Binational Emission Reduction Credit Trading
No. 11 Dynamics of Human-Environment Interactions
No. 12 Integrated Approach to Defining Particulate Matter Issues in the Paso del Norte Region

Contents

Foreword

The All-American Canal: An issue of contexts

Water and environment recognize no political boundaries. That's a maxim of environmentalism that preaches to erase borders so to properly manage natural resources. In the changing landscape of US-Mexico relations, water disagreements have consistently made life difficult for diplomats in Washington and Mexico City; the current debate over the proposed lining of the All-American Canal (AAC) represents the latest and one of the most complex controversies. Borders are foreign to natural systems; they demarcate views and management regimes, and impede the free flow of information from one side of the line to the other. By most measures, the Colorado River Delta Region is one of the most studied environments in the US-Mexico Border region. Yet, thorough knowledge of the system by both countries has not translated into a more relaxed, less apprehensive mode of diplomacy. (See page 24 for map.)

The Lining of the All-American Canal constitutes a comprehensive review of one more natural resource management issue on an international border. It is, however, by no means, a typical bilateral environmental issue on a typical international boundary. The US-Mexico Border is perhaps the only area in the world where a highly industrialized nation co-exists with a developing nation, jointly addressing the effects of their associated differences in income, culture, language, and religion. Moreover, the All-American Canal (AAC) issue is an atypical disagreement within an otherwise agreeable relationship between the two countries, one which has become increasingly cooperative in areas such as management of air quality, hazardous waste, and chemical emergency response and preparedness.

Lining the All-American Canal: Competition or Cooperation for the Water in the U.S.-Mexican Border?

Since the book's first publishing in Spanish (August 2004), negotiations between both governments have intensified, confirming that the collision of water and borders is neither new nor unfounded. Several authors in this book describe the tension over the AAC brewing since the late-1970s/early-1980s when the U.S. Department of the Interior decided that the State of California needed to cut its use of Colorado River water down to its legal apportionment of 4.4 million acre-feet per year (MAF/yr). The thirsty and thriving West and the historically over-apportioned Colorado River left the State of California and the U.S. Federal Government with apparently no alternative but to recapture the waters seeping under the All-American Canal. On October 18, 1983, the U.S. Government formally notified Mexico that those waters seeping underground and across the border were surface waters from the Colorado River apportioned to the U.S. under the 1944 U.S.-Mexico International Water Treaty.

But the debate over the All-American Canal seems not only to linger but deepen. A multitude of complex circumstances exacerbate the longstanding impasse in this negotiation between the two countries. At the top of the list is the growing recognition of just how politically intertwined water and environment issues have become. Additional complexity is added as new agencies from both governments have been brought to the table, raising the question whether age-old institutions and processes that were once considered dogma are still adequate. No longer can the purview of an institution like the International Boundary and Water Commission (IBWC) be considered all-encompassing when dealing with U.S.-Mexican water issues, nor can the 1944 U.S.-Mexico International Water Treaty be relied upon to resolve matters arising from increasing development and the shortage of water in the region.

In addition, as time passes, the list of stakeholders willing to take action appears to be growing. Along with both federal governments and their growing list of involved agencies, the State of California and Baja California have begun dialogues with each other as well as with their respective national governments. Civil groups North and South of the border have begun to collaborate on the issue, and environmental non-government groups continue to voice their concern over the effects of the piecemeal management that has taken hold of the Colorado River Delta region. Potential socio-economic impacts on both sides of the border have prompted still more groups to seek out-of-the-box alternatives for solving the conflict.

The water-environment conundrum

Water is a part of nature. But not all water issues are dealt with as part of the international environmental agenda. Historically, water issues between Mexico and the United States have been resolved under the jurisdiction of the International Boundary and Water Commission. But the agency needed a shift in culture to adapt to the changing times. Minute 306 of the IBWC recognized a growing sense that environmental issues could not be avoided when managing the Colorado River and was signed on December 12, 2000. Minute 306 reflected the emergent collaboration between environmental agencies across the border as well as scientific, academic and non-governmental organizations interested in preserving the "ecology affected by decreases in Colorado River flows in this reach and potential impacts to the habitat of fish, marine and wildlife species of concern to each country." Concurrently, Bi-National Technical Work Group IV was formed and charged with carrying out the intentions stated in Minute 306.

But the inertia of bureaucracies is hard to break. Water and environment quickly became two separate issues again for the IBWC as political pressures over the lining of the AAC heightened. This, in spite of studies describing how ending seepage from the All-American Canal would result in loss of wetland habitat in the Andrade Mesa that would impact the endangered Yuma Clapper Rail (Rallus longirostis yumanensis) along with about one hundred migratory species in the Pacific Flyway (Hinojosa-Huerta et al, 2002). The U.S.-Mexico bureaucratic milieu formed a fifth work group to discuss the AAC, rather than assigning the discussions to the already environmentally chartered Group IV. The U.S. Bureau of Reclamation (USBOR), the lead water agency for the U.S. within the group, drafted a Scope of Work (SOW) which called for two subgroups, one to discuss hydraulic issues pertaining to the lining of the AAC and another to discuss groundwater issues. Through the IBWC, Mexico requested that the impact on the Andrade Mesa wetlands be part of the SOW for Group V, but progress stalled as the U.S.-Section refused to include environmental issues as part of the AAC discussions. Water issues continued to be negotiated within a narrow political perspective, as if the environmental consequences could be placed on hold or wished away.

Since then, discussions on the AAC have again spilled over into the environmental arena. On September 1st 2004, Mexico's Secretary of the Environment and Natural Resources addressed a letter to the U.S. Interior Secretary, marking the first time that environmental agencies outside the Foreign Office (Department of State and Secretaría de Relaciones Exteriores) had raised an issue concerning waters along the boundary with its international counterpart. The letter has forced bureaucracies to acknowledge the interconnectedness of nature, leaving agency jurisdiction as the hindrance next in line.

Nature parceled as agency purviews

In both countries, water agencies only address water issues without concern for environmental impact, and agencies that address the environment may only tend to environmental matters, leaving water issues to the water agencies. Wetlands and wildlife in the U.S. are the responsibility of the U.S. Fish and Wildlife Service (USFWS), which enjoys a vigorous working relationship with Mexico's National Institute of Ecology (INE). But the USFWS will not discuss mitigation and cooperation in the Andrade Mesa Wetlands with INE given the area's association with the All-American Canal. In order for USFWS to address possible restoration in the Andrade wetlands or any part of the Colorado River Delta, conversations must be concurrent with the U.S. Bureau of Reclamation, given BOR's lead on water management issues. But the USBOR has to date only held discussions on Colorado River water infrastructure with the Mexican National Water Commission (CNA) and has rejected meeting with administrative and policy level representatives from SEMARNAT proposed as members for Work Group V.

Ultimately, all U.S. agencies placed AAC discussions on hiatus, as legal action against the U.S. Government was filed in July 2005 by civil groups in Mexicali and Imperial Valley over the procedures followed in the environmental impact studies. The Department of Justice (DOJ) is in charge of responding to legal suits against the government, and subsequently became the lead agency on all AAC talks.

SEMARNAT has altogether addressed three letters to its counterpart DOI Secretary Gale Norton to propose a bilateral task force to discuss the lining of the AAC and its environmental impacts. The thinking behind this Mexican initiative was that raising the discussion to cabinet level members would enable a comprehensive view of the problem and spark a coordinated effort from both bureaucracies involving technical experts as well as policy level personnel, all for the benefit of the border environment. The DOI, however, responded that the issue should be taken up with the IBWC, given that this agency had the lead in consulting Mexico on bilateral water issues. Bureaucratic purview prevailed, and all negotiations stalled.

For all of SEMARNAT's efforts to engage its counterpart DOI in bilateral discussions on the AAC, at the end of the day, Mexico's Foreign Affairs Secretariat (SRE) is the agency in charge of bilateral relations for Mexico. The same holds true in the U.S., where the Department Of State takes the lead. The formal role of environmental agencies in a diplomatic context is to advise the Foreign Office on environmental matters. SEMARNAT, DOI, and EPA, for that matter, are the technical/scientific agencies that provide substance on international issues, not policy.

This is a fact of diplomatic life and a context shaping the negotiations on the AAC that cannot be overstated. The U.S. and Mexico are simply too interdependent in commercial, political, and social terms to allow loose ends to get in the way of their bilateral agenda. At the center of the complex relations between the U.S. and Mexico lies immigration, border security, and trade, not the All-American Canal. In that sense, it is to be expected that the intensity of talks reflect the mood and disposition of the relations stemming from other topics in the bilateral agenda. One example of this might be the case of Mexico's water debt in the Texas-Chihuahua border, finally covered in 2005, thus awarding Mexico with enough moral capital to once more state its objections over the AAC and raise the issue repeatedly before numerous officials in the U.S. Government.

In the course of the last year, conversations on the AAC between officials from both countries have multiplied considerably. The list is long and includes the Commissioners at the IBWC, SEMARNAT and DOI officials, SEMARNAT and EPA officials to consult agency purview, the Mexican Ambassador in Washington with numerous officials at the DOS, and Mexican Secretary of Foreign Affairs, Luis Ernesto Derbez, with U.S. Secretary of State, Condoleezza Rice. Even Presidents Bush and Fox engaged in a conversation on the All-American Canal at Crawford Ranch in March 2005, a dialogue which sparked a series of meetings in a multi-agency setting in Washington that provided hope but as of yet, no progress.

This diplomatic context, however, does not preclude existing contacts and ongoing cooperation between environmental agencies that result from history and thematic proximity. In that sense, the environmental agencies play a dual role carefully balancing their duties within their own government with the relations they hold with their international counterparts.

This is explicitly the case of the *Agreement between U.S.-Mexico on Cooperation for the Protection and Improvement of the Environment in the Border Area,* a.k.a. The La Paz Agreement, signed in 1983 "to establish the basis for cooperation between the Parties for protection, improvement and conservation of the environment" in the border area (Article 1), framing the parties' commitment to assess, within their legal framework, "projects that have significant impacts on the environment of the border area [so] that appropriate measures be considered to avoid or mitigate adverse environmental effects" (Article 7).

The Environmental Protection Agency (EPA) was entrusted with the task of administering the agreement on behalf of the U.S. and the Secretaría de Desarrollo Urbano y Ecología, a predecessor to SEMARNAT, for Mexico. This designation formalized a direct cross-border relationship between the agencies, which enabled a string of border environment cooperation programs, Border 2012 being the latest. EPA and SEMARNAT have also become main players in evolving NAFTA environment institutions, namely the Border Environment Cooperation Commission (BECC) and the North American Development Bank (NADBANK).

But, The La Paz Agreement, signed by Presidents Reagan and De la Madrid, was never submitted to Congress for ratification. So, implementing the agreement itself to its full extent, let alone enforcing a specific article in it for the purpose of resolving a bilateral conflict, remains a complicated undertaking within the national bureaucracies. Even though EPA and SEMARNAT are in charge of its compliance, agency turf is a sensitive issue in government circles, and agency officials transgress this rule at the risk of waging political battles within their administrations. Be it historical priorities or a reality born of practice, the La Paz Agreement has contributed little to improving conditions for greater cooperation in cross-border negotiations.

*The AAC: not **just** a federal issue*

The most significant action of late by a non-federal stakeholder to the AAC is the lawsuit filed by three civil society groups: the Council on Economic Development of Mexicali (CDEM), the Citizens United for Resources and the Environment (CURE), a group based in the Imperial Valley and Desert Citizens Against Pollution. A petition and notice letter was first filed in mid-May and a lawsuit in July 2005 questioning DOI's compliance of National Environmental Policy Act (NEPA) procedures regarding the Environmental Impact Statement/ Environmental Impact Review on the AAC lining project. It essentially brought discussions between U.S. and Mexican federal agencies to a halt, as the U.S. Government decided that only DOJ lawyers could speak on the subject. Other U.S. non-governmental organizations, the Sonoran Institute, Defenders of Wildlife and Environmental Defense among others, have addressed several written communications to Secretaries Rice and Norton and other top officials in the federal Government to bring their attention to studies showing significant impacts to critical wetland habitat in the U.S. and Mexico. The main concerns outlined by these groups include not only the environmental impact of the canal lining, per se, but also complications to settlement obligations to the San Luis Rey tribe, causing a possible disruption of Colorado River-related operations. Several tribes from the Colorado River Indian Tribes and the Ten Tribes Partnership, notably the

Cocopah and he Quechan Indians, have joined the chorus of non-federal players, voicing concerns over several aspects of the Colorado River Water Delivery Agreement, ranging from omissions by the BOR's Environmental Impact process to the relegation of the seniority of their water rights.

Breaking the impasse

If there's ever been a natural resource management issue where government and diplomatic institutions along the U.S.-Mexico Border must rise to a place of leadership it's the AAC. The livelihood of economic players and of indigenous groups that have straddled the border since time immemorial is at stake, not to mention the future of an ecosystem with unique characteristics. Given the great effort invested on cooperation by both federal governments, the state governments along the border, the academic community, the NGO community, and society in general, the matter at hand now is how Mexico and the United States will employ their cooperative outlook on border environment issues to solve this spiny topic.

A major adjustment to the current terms of exchange between both countries is needed in order to break the impasse on the AAC. Bureaucracies must conform their approach to the goals staked; the ends should justify the means, and not the opposite. Nature requires a joint not a parceled perspective. Therefore, a system of discussion must take hold erasing the artificial boundaries of agency purview, so that proper management of a system as complex as nature can be successful.

Two notable tools of U.S.-Mexican diplomacy hold promise. First and foremost is the La Paz Agreement, which has been grossly underutilized. This twenty-plus-year old document once enshrined a basic set of rules by which both countries could envision a cordial and respectful co-existence along their common border and a sustainable future for their communities. The essentials of the La Paz Agreement boil down to two factors: communication and bilateralism. Communication requires the free exchange of information and a congruent dialogue that leads to solutions of problems. Bilateralism implies the conviction that the border area is a shared responsibility and that solutions to problems of mutual concern must be determined jointly. Reciprocity is a prime tenet of the accord, and a value to which both countries committed to in the agreement.

The Transboundary Environmental Impact Assessment (TEIA) was mechanism specifically devised by the environmental agencies of Canada, the United States and Mexico to streamline solutions to cross-border environmental problems. The TEIA has hit a dead end for now, as differences in the legal frameworks of Mexico and the U.S. have prevented its implementation. Both mechanisms, the La Paz Agreement and the TEIA will cease to exist as tools for diplomacy unless both countries exhibit a renewed determination and a political will to revive them.

Can't fool Mother Nature

The Colorado River Delta region and the human infrastructure created for its use may be the poster child of hybrid natural-manipulated systems. For over a century, the Colorado River system has witnessed a continual action-reaction mechanism, and the results to the natural ecology have been significant. Manipulation of the Colorado is responsible today for the creation of systems perceived as natural, like the Salton Sea, and for the demise of others, most notably a rich and highly productive estuary in the Gulf of Santa Clara. Many species, such as the totoaba *(Cynoscion macdonaldi)*, have been driven to virtual extinction, and others, currently listed under the Endangered Species Act, are further impacted by the river's management.

It's a repeated truism that nothing goes unnoticed within a system: when one element is altered, another must adjust. John Muir once said that "When we try to pick out anything by itself, we find it hitched to everything else in the Universe." Hence, the lining of the All-American Canal will undoubtedly fail in fooling Mother Nature. If the lining of the Canal does occur, impacts will accrue to the already altered Colorado River Delta Region.

Though all stakeholders involved appear to realize this inexorable truth, two questions are left for future debate. First, how does this fact of nature play itself out in the diplomatic and political realm, given the multiplicity of stakeholders and stakeholder interests involved in the Delta? And second, assuming that further aggressions are tolled on the natural system, how long will the system stand without collapse?

Carlos A. de la Parra Rentería

Preface

This volume on the All-American Canal is a most welcome addition to the literature on transboundary water resources, U.S.-Mexican relations, and the role of laws, institutions, and diplomacy in contemporary international affairs.

Appearances aside, this volume is about much more than one instance of binational conflict. The All-American Canal story speaks directly to the often-troubled history of relations concerning shared water resources and the present strain on U.S.-Mexican relations. While the accounts of events, failures in negotiation, and present inertia pertaining to the problem are frustrating and even discouraging, this book highlights some reasons for hope and even optimism.

Embedded in the very name "All-American Canal" is a clue to the basic issue at stake. Unilateralism is at odds with geographical and hydrological reality in the river basins and aquifers that straddle international boundaries. To behave as if water were ice cubes that can be separated and dumped into different glasses held individually by the parties involved leads to problems all around. In dozens of locations on the very long U.S.-Mexican border, how one nation handles water on its side of the border very much affects the welfare of the other. Saline discharges from irrigation in Arizona degrades the quality of Colorado River water flowing into Mexico and even precipitated an international crisis in the early 1970s. Present day pumping by the Chihuahuan municipio of Ciudad Juárez from an aquifer shared with the Texas city of El Paso threatens the economies of the binational region.

From the very beginning the All-American Canal was an inappropriate answer to a misconceived problem. Action was precipitated by a breach in the engineering works related to the irrigation diversions and canals partly in Mexican territory, but largely serving U.S. investors. For nearly two years the Colorado River poured into low lying areas, flooding farms and cities including Mexicali, Baja California, and filling what is now known as the Salton Sea in

California. Poorly constructed works and inadequate and short-sighted private investment were mainly to blame. But, in authorizing and funding the project, federal agency officials and members of the U.S. Congress argued that it served national security. Rather than framing the issue appropriately as a joint management problem, the fault was placed on the location of the Alamos canal running through Mexico but serving irrigators in both the Mexicali and Imperial Valleys, in Baja California and California, respectively. Sovereign national power was portrayed as essential to security in an area where limited national territorial sovereignty would have been more appropriate. Unilateralism and "going it alone" in relation to the All-American Canal set an unfortunate precedent, which has been a strategy followed by the United States far too frequently with its friends and allies.

Progress toward binational, joint water management has occurred in fits and starts and is far from sufficient. The International Boundary and Water Commission-Comisión Internacional de Límites y Aguas (IBWC-CILA) has more than 60 years of experience in managing border water and is sometimes celebrated as a model international institution. However, in practice, the institution operates to protect sovereignty when separate national interests are in conflict with open collaboration. Each section is organized separately with its own engineers and sources of data that are not always shared or available to the public. The La Paz Agreement, signed in 1983, requires the United States and Mexico to cooperate in resolving environmental problems of mutual concern and led to the establishment of a number of task forces and working groups with membership from federal, state, and local authorities from both nations. The La Paz Agreement has been more a symbol and a framework than a blueprint for action. The Border Environmental Cooperation Commission (BECC), established as part of the North American Free Trade Agreement (NAFTA), is laudably inclusive in terms of membership embracing the environmental agencies in both countries as well as state and public members. However, its portfolio is limited to certifying planning and construction of infrastructure projects. The behavior of the United States in failing to consult with Mexican authorities prior to the decision to alter the All-American Canal is not helpful in advancing holistic, integrated transboundary water management.

Preface

One of the contributions promised in the publication of this book is to raise the visibility of this terribly important problem and to inform the discourse. While the plan to line the All-American Canal in 2006 gives this story immediacy, it is anything but new. It has been part of regional and state agency planning for nearly 25 years. Both in conference speeches and in his writing, the late Albert E. Utton raised a red flag about the topic among members of the water resources legal and academic communities. However, in the halls of power in Washington D.C., this issue has, until very reently, generally been ignored. Lack of attention can be partly explained by crises such as illegal drug trafficking and undocumented immigration, which tend to eclipse all else. More is involved however, and it is generally the case that Mexico takes its relationship with the United States much more seriously than the United States takes Mexico. Matters of great importance to Mexico in the bilateral relationship seldom make the U.S. national agenda unless Mexico somehow gains some leverage through appeals to the international community or holds out, until its issues are addressed, on some arrangement the United States badly wants. It must be remembered that the United States avoided water allocation treaties related to either the Colorado River or the Río Grande until the Roosevelt Administration made such a treaty part of the concession to Mexico for joining allied forces in World War II. It is not naive to believe that really good books such as this one have some power to create an attentive audience.

The story of the All-American Canal presented here highlights the failure of existing bilateral institutions to resolve water problems between the United States and Mexico. While IBWC-CILA has very broad authority, it has not used its authority to provide joint consideration of groundwater apportionment. Ultimately the two nations must agree on a groundwater treaty, but each year the numbers of groundwater users increase, the levels of shared aquifers decline, and local interests using groundwater become more recalcitrant, making the negotiation of fair and equitable apportionment more difficult. While a disappointment to some, the new international environmental institutions created by NAFTA present new forums and provide some new bargaining resources, including new infrastructure and funding to mitigate and compensate Mexico for harm that may be suffered from lining the new All-American Canal. Even so, the responsiveness of these institutions is very limited.

The real issue raised here is whether the enormous population growth experienced on both sides of the border is sustainable in the face of the physical limitations of the regions' rivers and aquifers. The All-American Canal is as much about the issue of growth as it is about international allocation of water. Urban users, particularly the City of San Diego, California, will win a greater share of the Colorado River at the expense of agricultural interests in the Imperial Valley. To many this reflects a movement of water to higher and more efficient uses through markets. It must be remembered, however, that unlike alfalfa and cotton fields that can be fallowed in times of drought, municipal and industrial users cannot simply turn off the faucets. It may well be that the collision between the imperative of drier climate predicted for the border region and the actuality of unabated population growth (which is facilitated by the new All-American Canal) will be greater than any previous challenge.

Finally, this volume is laudable for demonstrating the great strides made in scholarship on the shared natural resources among the United States and Mexico. The norm of the past has been a thin literature coming from mainly U.S. scholars in border-state universities with only a trickle of academic writing from Mexicans presenting a Mexican perspective. Over the past several decades, academic and research institutions located near the Mexican border have improved dramatically in quality, and there is an outpouring of books and articles on a whole range of common resource issues. While there is great asymmetry in the national power relationship between the two nations, there is something approaching parity in relevant policy scholarship. This book will inform relevant policymakers and citizens in Mexico and the United States, and to the extent that knowledge is power, it will forge a powerful negotiation force for both.

Helen Ingram
Warmington Endowed Chair, Social Ecology, University of California, Irvine
Distinguished Research Fellow, Southwest Center, University of Arizona

Introduction

The evolution, growth, and transformation of human societies could not be explained but for access to water. As such, the control of water as a resource and the economic development achieved throughout the 20th century are inseparable and interrelated elements.

Control of water resources has become a source of discord and has increased the possibility of conflicts between nations. Peter Gleick recounted the conflicts between 1503 and 2000 that centered on water and found they have occurred with greater frequency since the 1940s. In his study, it was clear that the U.S.-Mexican border region has seen the least amount of conflict over water throughout history.[1] But now there is an increased risk of conflict over water access due to greater demand in the face of scarcity, which is a consequence of global climate change as well as the overexploitation of traditional supply sources.

In the U.S.-Mexican border region, where supply is low and demand high, the risk of a conflict over water is great. Yet, there have been no significant conflicts in the past over its distribution. This makes the region an excellent case study for evaluating the strategies the two countries have used to settle their disputes over access and control of water—that is, a case study on the political will to reach rational and peaceful agreements on the distribution of a scarce resource.

Although scholars have found inefficiencies in the law that regulates the distribution of water, the law itself is not a rigid, static, or unchanging one. Rather, its design allows for subsequent agreements, called Minutes, to be incorporated into the law to resolve specific issues. This makes it possible to adapt to the changes in management of water on either side of the border. As well, new agreements have been forged between the United States and Mexico that broaden understanding and cooperation in the care and management of the resources located on the shared border. In 1983, the

United States and Mexico signed the Agreement on Cooperation for the Protection and Improvement of the Environment in the Border Area. Referred to as the La Paz Agreement, it calls for binational cooperation to reduce pollution and restore the environment throughout the border region, defined as the area situated 100 kilometers on either side of the U.S.-Mexican boundary. As part of the North American Free Trade Agreement between the United States, Mexico, and Canada, the North American Agreement on Environmental Cooperation was signed in response to public calls in the United States and Canada for environmental protection. New institutions were created and charged with enforcing the agreement and administering the programs supporting environmental projects.

There are, though, important challenges for water use and management in the region. Growing demand for the resource and its limited availability are the factors determining new water policy. Until recently, the only concern of water-management agencies had been to obtain greater volumes of water to satisfy the projections of future demand. Although, generally speaking, this pattern continues, different potential policies are now being discussed, such as savings, conservation, treatment, and reuse, which taken together represent an important qualitative change and create hope that water culture will evolve in terms of water-use practices and institutional management models.

Concerns abound on both sides of the border about environmental risks from exploitation of water sources as well as the need to restore the ecosystems in the watersheds from which water resources are obtained. These ecosystems are have been or currently are being negatively impacted by the lack of minimum, good quality water flow. They are the natural environment upon which a vast array of species depends.

The U.S.-Mexican border faces a bleak future in terms of the distribution of water from shared watersheds. No additional water is available, and what one side somehow obtains will be at the expense of the other or of the environment itself, which could translate into a severe or even irreversible conflict or environmental catastrophe. Nevertheless, there is the historical precedent of the United States and Mexico working together to resolve their controversies and conflicts. The institutional framework they have created to deal with potential conflicts focuses on cooperation.[2] Logic dictates that this attitude will continue, given the growing integration of the economic interests of both countries and the dynamic that the border region itself represents.

This context is the framework within which this volume analyzes the conflict over the lining of the All-American Canal. This canal runs parallel to the international border and delivers Colorado River water from the Imperial Dam to the agricultural areas of the Imperial Valley in California. Because it is not lined with concrete and instead is dug from porous sand, for decades it has fed the Mexicali aquifer with the water that seeps from it, and in turn has provided Mexicali Valley growers with water to irrigate their lands.

In 1988, the U.S. government approved the canal-lining project over the manifest opposition of the Mexican government, which balked because of the negative impact it would have on the Mexicali Valley growers. Although the project has been postponed several times for various reasons by the United States, in recent years resources have been approved for both the lining and the conveyance of the recovered water to the urban region of San Diego, California.[3] California's governor at the end of September 2003 granted an extension for the project from 2006 to 2008. But, once the All-American and Coachella Canals are lined, the water will be sent to San Diego for the next 110 years, per an agreement reached between the water authorities of Los Angeles and San Diego.

In 2005, at a trilateral meeting on security between the governments of the partner countries to the North American Free Trade Agreement, the Mexican government raised the issue of lining the AAC, but only after the government of Texas raised the issue of the historic water debt from the Rio Grande watershed. Since that meeting, both governments have been negotiating a solution to their differences over the AAC lining project.

The objective of this volume is to provide a broad vision of the context within which the decision to line the canal was made and the implications the lining has on the Mexicali Valley and the state of Baja California, both of which are experiencing accelerated growth in water demand and whose only stable supply is Colorado River water. Possibilities for a solution to the controversy and conflict surrounding the lining of the AAC are presented and explored.

They offer a different direction than proposals floated by the governments and take advantage of the existing institutional framework of cooperation. They incorporate new visions and models for water resource management in a crossborder context. In this sense, this volume is a compilation of works from different disciplinary perspectives that provide and analyze the region's biophysical and ecological information, quantify and address the quality of the water, and outline the probable impacts on the Mexicali Valley and, to a lesser degree, Baja California.

This volume is a collective project resulting from the academic collaboration of experts concerned about one of the issues that will surely play an ever more important role in the binational agenda—water management. The multidisciplinary approach to the issue is one of the enduring benefits of this work and will result in a more integrated vision of the problem and the binational management of the resource.

The lining of the AAC is a microcosm of other problems with shared natural resource management along the border between the United States and Mexico. Here, the optimum advantage and conservation of resources demand, above all, imagination. They will engender cooperation and the participation of the neighboring communities in an effort to overcome the asymmetries and points of view anchored in unilateralism, as is the status quo.

Some of these works have their roots in projects sponsored and completed by investigators at El Colegio de la Frontera Norte (COLEF) and the Centro de Investigación Científica y Estudios Superiores de Ensenada (CICESE). Others are the result of work carried out in other institutions of the region, such as Universidad Autónoma de Baja California (UABC) and Instituto Politécnico Nacional in Northwest Mexico. Still others are the result of long histories of study of the institutional problems in crossborder water management, as is the case for Stephen P. Mumme and Donna Lybecker of Colorado State University; María Rosa García Acevedo of California State University, Northridge; and Francisco Zamora and Peter Culp of the Sonoran Institute. Zamora and Culp, along with Osvel Hinojosa of Pronatura, kindly accepted the invitation to contribute the chapter on the environmental impact of lining the AAC, which greatly enriches this English edition.

This volume seeks to contribute, inasmuch as possible, to the understanding of the issue and to help find a satisfactory solution to it. The authors explore everything from classically state-centric solutions to the role of non-traditional stakeholders, such as Congress, local governments, and even the possible links between crossborder stakeholders.

This volume is organized into three sections. The first deals with aspects related to the biophysical and environmental characteristics of the region. Here, Francisco Raúl Venegas summarizes the most notable biophysical characteristics of the region, where the Colorado River and its delta come together yet are fragmented by their belonging to different political units. Francisco Zamora, Peter Culp, and Osvel Hinojosa list the biodiversity that characterizes the area where the AAC is located, quantifying and typifying the damage the environment is expected to sustain upon the project's implementation. They also establish the limitations of the environmental impact statements performed by the United States government to gain approval of the project.

Jaime Herrera Barrientos, with a team of colleagues, compares the Mexicali aquifer characteristics with the hydrology of the AAC to establish the relationship between them and then determine what affect losing the seepage from the AAC will have on water quantity and quality. In the third work in this section, Gerardo García Saillé, along with Ángel López and José Antonio Navarro Urbina, study the existing relationship between the AAC and the Mexicali aquifer's geohydrology to determine how the loss of recharge from the canal's seepage will affect on the aquifer's water quality.

The second section covers the sociopolitical problems in the environment in which the AAC's lining decisions have been made. Fernando Medina applies an historical perspective to the visions that have governed water-related decisions in the American Southwest, and in particular in the border region, pointing out the adjustments that have taken place between the river-bordering states as a central element in understanding the water-policy context of the region. María Rosa García Acevedo addresses the vision with which the political discourse surrounding water has taken place in the United States, and the fact that the resource management policy actions are geared toward making the countries either winners or losers of a shared resource. José Luis Castro analyzes the most likely scenarios in water availability and management in Baja California cities in the long run, concluding that it is the region's urban environment that will feel the effects of lining the AAC, and that the responsible authorities face considerable challenges that require timely decisions to satisfy a growing demand, now and in the future.

Chapters in the third section address the alternatives that both the U.S. and Mexican governments could follow to find a mutually agreeable solution, building on a spirit of cooperation rather than competition, thus avoiding conflict over lining the canal and the damage it will cause to Mexicali's economy. Instead, these chapters suggest, the governments could set an enduring precedent that could be the basis for the resolution of other binational conflicts over water.

Stephen Mumme and Donna Lybecker focus their analysis of the differences between the United States and Mexico on lining the AAC and its negative impact on Mexico using the theoretical perspectives of international law as a starting point. They also use the theory of common resources and game theory to establish the possibility that the parties can arrive at an agreement based on cooperation and exchange, overcoming the zero-sum position under which they have been operating until now. These authors suggest that recent developments in international law theory open the door to considering that this case is a matter of that nature. In this regard, Mexico would have to avail itself of international tribunals to pressure the United States and force a bilateral agreement based on the principles of exchange and mutual concessions.

Vicente Sánchez analyzes the position Mexico has maintained and the negotiations between the two governments. He presents the possibility that the two countries could reach a cooperative agreement, considering that historically they have built an institutional framework to do so in the past. However, this will require that the Mexican government abandon the low profile it has maintained on this issue and assume a defined position and a negotiating strategy with an integral vision that goes beyond the issue of lining the AAC and considers other environmental elements that are part of the geographical unit where the canal is located.

Finally, Alfonso Cortez Lara deals with water availability in the Mexicali Valley and the impact of lining the AAC. The author posits that arriving at a mutually advantageous solution requires a change in the principles upon which the negotiation has been established. Until now a market equilibrium criterium has prevailed, while criterium of equilibrium between the parties in the game has been absent. This demands greater involvement of various agents on both sides in building a solution, as well as the establishment of an information base for both sides of the border with the same criteria and methodology.

The editor thanks the Southwest Consortium for Environmental Research and Policy (SCERP) for its support in the translation and editing of this volume, Rick Van Schoik and Dr. Paul Ganster for the support and the kindness with which they initiated the publication of this work in English, the authorities at COLEF and Plaza & Valdez Publishers for granting the editorial rights and allowing the publication of this book in English, and Bertha Hernández for her patience and skill as executive liaison in developing this final product. Thanks also goes to colleagues at the Sonoran Institute and Pronatura for their collaboration.

ENDNOTES

1 According to Gleick, the only conflict recorded in the region occurred between California and Arizona in 1935, when Arizona mobilized National Guard units to its border with California in protest of the construction of Parker Dam, which would divert water from the Colorado River to California.

2 The International Boundary and Water Commission (IBWC) is the binational agency charged with implementing the agreements between the United States and Mexico related to territorial limits and ordering the water assignments agreed to in the 1944 Water Treaty, which apportioned the volumes of water from the Rio Grande (Río Bravo), Colorado River and Tijuana River. IBWC also has remediation attributions along the border. In recent years the institutional framework for environmental protection has been strengthened. In 1983, the La Paz Agreement was signed relative to cooperation for environmental remediation on the border. Later, as

part of the North American Free Trade Agreement (NAFTA), the Border Environmental Cooperation Commission (BECC) was established, as were the North American Development Bank (NADBank) and the Border XXI and Border 2012 programs, through which binational cooperative efforts on environmental objectives are funneled.

[3] This is according to articles in the *Los Angeles Times* and *San Diego Union Tribune.*

I

Physical and Biological Features of the Colorado River

Francisco Raul Venegas Cardoso

INTRODUCTION

A great deal of information on the Colorado River Watershed is available in works by Sykes (1937) during the first quarter of the 20th century and in descriptions by early settlers while traveling across a vast desert, limited by what was then known as the Sea of Bermejo. These first Europeans discovered, among other marvels, the majestic and imposing Colorado River forming what are known today as the Imperial and Mexicali Valleys. They learned that the sediments constantly brought by the Colorado River were being reshaped by coastal processes. There was a natural channel in the Laguna Salada (Salt Lake), between Sierra Cucapah and Sierra de Juarez, that would quite possibly have been navigable. And there was a vast extension of land constantly being watered and fertilized—the Colorado River Delta. Amidst this great desert environment, the Colorado River Delta rises as an oasis surrounded by uniquely shaped plant species that owe their form to living under extreme drought conditions. A wide variety of animal life also exists there.

This chapter provides an overview of the physical and biological features of the region. It lays out chronologically the past and present history of the Colorado River, it then continues with a description of its current state. Finally, it ends with a description of the trend toward a potential recovery of the flows and significant flood waters that existed prior to the construction of dams in the United States.

Physical Features

Geology

The ancient history of the formation of the Baja California peninsula, as it is known today, dates back 135 million years to the Mesozoic era. It began with an active system of faults known today as the San Andreas and its parallel branches, the Elsinore and San Jacinto faults. The high activity of this system caused the detachment of an immense block that formed what is now known as the Salton Trough (Gastil, Phillip, and Allison 1975; Singer 1998). After this deep depression formed, an intensive erosion process filled it with sediment of gravels and sands from the San Bernardino Mountains.

This transportation and sedimentation process lasted approximately 70 million years and today those sands and gravels have partially consolidated into sandstone and conglomerates (Muffler and Roe 1968, Singer 1998). The depth of the sediment layers deposited over several million years ranges from 2,060 meters (m) in Indio to nearly 6,000 m in the case of deposits located at the current U.S.-Mexican border. The Mexicali Valley is also part of this depression, as evidenced by a stratigraphic column of the different sedimentation events at Cerro Prieto. Their possible ages range from the Mesozoic to recent times and depth ranges from 2,000 m to 2,500 m (Puente and de la Peña 1978).

At this time, the Baja California peninsula had not yet detached from the continental mass (this process did not begin until the Miocene era, approximately 5 million years ago). The San Andreas fault system became active once more and began forming the Proto-Gulf of California as well as causing the separation of the Baja California peninsula (Gastil, Phillip, and Allison 1975). Other authors suggest that the main San Andreas fault became active again 4 million years ago (Elders, et al. 1972, Singer 1998). Regardless of which detachment hypothesis is most accepted, the separation continues today and the peninsula will eventually detach from the continental shelf. As Elders, et al. (1972) indicate, this separation is occurring at a rate of 5 centimeters per year.

The Salton Trough encompasses the Imperial and Mexicali Valleys and is part of the San Andreas fault system, one of the most active in the world (Lira 1994) and a result of the collision of the Pacific and North American plates. The high seismicity of the entire region is attributed to this system. Because the region as a whole has great economic importance due to its geothermal energy fields and its significant agricultural potential, tectonic, seismic, geophysical, and sedimentary research on it is continuous. This work indicates that as this large mass of land detached from the continental shelf, the marine waters of the ancient proto-gulf reached as far as the northern edge of what is known today as the Salton Sea, approximately 260 kilometers (km) north of the U.S.-Mexican border. After the formation of the proto-gulf and the detachment and expansion of the current Gulf of California, the Colorado River began its great task of conveyance, sediment discharge, and leveling of this deep depression.

Singer (1998) reports that during the glacial periods of the Pleistocene, which occurred in the last 2 million years, the Colorado River has been depositing its sediments, building the delta into its current shape as it flows into the Gulf of California. There were four glacial periods during the Pleistocene, each with their respective interglacial period (Derruau 1970). During those vast expanses of time, the river ran with abundant floodwaters. During interglacial periods, however, the amount of water flowing in the river and its accompanying sediment load diminished significantly, leading to droughts and intense evaporation that at times left the riverbed dry and barren.

According to Singer (1998), just more than 1 million years ago the already-formed Colorado River Delta served as a barrier that kept the waters and sediments of the river from continuing on to the sea. Instead, it filled the deep depression known as the Salton Sink, where the Imperial and Mexicali Valleys now lie, giving birth to Lake Cahuilla. This ancient body of water was possibly one of the largest freshwater lakes—its surface area may have reached 3,320 km2, perhaps six times the size of today's Salton Sea, and expanded even into part of the Mexicali Valley. This ancient lake has left several clues about its existence, such as in the beach lines from Indio to Cerro Prieto in the Mexicali Valley, where coastline watermarks reach as high as 12 m. According to existing data, this has not occurred since the end of the last glacial period approximately 21,000 years ago and the end of the Pleistocene 12,000 years ago (Ortega 1995).

Evidence of the lake exists from just 1,200 years ago; this evidence also suggests that by 1600 AD the lake had completely disappeared. However, testimonials collected during the Spanish explorations about the existence of an ancient lake indicate the natives of the region had used it in ages past. The formation and disappearance of this ancient lake demonstrates how dynamic the Colorado River was. At times it would convey a large amount of sediment, causing the formation of banks or dams that diverted its flow or changed its course and created bodies of water that were constantly fed. At other times, the flow would stop and these bodies of water would begin to dry out from evaporation and the lack of a fresh water supply. Today, the Colorado River Delta covers more than 7,700 km and is located completely within what is known as the Mexicali Valley.

Seismicity

The chronology of Baja California's seismic history was first recorded during the early Spanish explorations. However, there are many information gaps within this short period and several earthquakes occurred for which there are no records (Molina 1991). Molina provides eye-witness accounts and contributes chronology and possible magnitude data for earthquakes recorded between 1776 and 1988. Lira (1994) reports that the first seismographic network was established in 1963 by the Universidad Nacional Autónoma de México (UNAM), the California Institute of Technology (Caltech), and the Centro de Investigación Científica y de Educación Superior de Ensenada (CICESE). A second network was established between

1977 and 1978, and a year later the Northwestern Seismic Network, allowing greater coverage and the capability of obtaining more data on this highly seismically active region.

Lira also presents a timeline of the frequent seismic activity between 1973 and 1993, making reference to the exact location, magnitude, and fault system associated with the quake. He concludes that the Cerro Prieto and Imperial faults are responsible for most of the quakes recorded during this period with magnitudes of between 1 and 6.6 on the Richter scale.

González (1990) traces the Imperial Fault and the rifts it is causing at Ejido Saltillo. The fault has exhibited a drop of the western block and rise of the eastern block, with a 1 m differential over the last 13 years. He concludes that a small, broad valley is currently being formed as a result of Imperial fault activity.

The Colorado River

The age of the Colorado River is still uncertain. Singer (1998) suggests that the lifting of the Colorado Plateau began seven million years ago, thus modifying its drainage. The new uplift of the plateau led to the formation of a drainage network with small creeks toward the southern end of the plateau, which gave new life to the ancient Colorado River. The renewed river found old courses, remaking its drainage pattern, cutting and forming the famous Colorado Canyon as it approached its final destination, the Salton Trough. Once it could no longer discharge its waters at this depression, it unloaded its sediment into the already-formed Gulf of California. Evidence of these events can be found in the sediment deposits correlated to the ages and types of rock, as well as fossils located in the formations aged approximately 11.8 million years. This leads to many authors' theories that the river is between seven million years and ten million years old.

Although its age may not be, it is certain that the river, from ancient times until prior to its being controlled by enormous dams, had always been a river with a sizable sediment load. It was composed of sands, gravels, silts, and clays (Muffler and Roe 1968, Singer 1998, and Puente and de la Peña 1978) that are deposited and divergent along today's Colorado River Delta.

The flow of the Colorado River has been fully controlled since the construction of Hoover Dam in 1935 and Glenn Canyon Dam, which filled between 1963 and 1980. The river's annual average flow totals 15.4 million acre-feet (MAF); however, some records indicate that during intense snow seasons in the upper basin, the volume may have increased to as much as 24.3 million acre-feet per year (MAF/y). Mexico, through the water treaty it signed with the United States in 1944, receives an allocation of 1.5 MAF/y.

Hydrologic Region No. 7, Colorado River

The Colorado River begins its flow in the United States and drains into the Gulf of California. The portion of the river managed by Mexico begins at Morelos Dam and serves the border between the states of Baja California and Sonora until it meets the Sonora-Pacifico rail line. According to the Hydrological Region classification of Instituto Nacional de Estadística Geografía e Informática (INEGI, in English National Institute of Statistics, Geography and Information), the Mexicali Valley lies within Region 7, Sub-Region 1, Río Colorado. Because the river runs on both sides of the border, all hydrographic observations about its flow are made jointly by the International Boundary and Water Commission (IBWC) in the United States and Comisión Internacional de Límites y Aguas (CILA) in Mexico.

The Mexican portion of the river is located between latitudes 31°45' and 32°40' north and longitudes 114°30' and 115°40' west. It flows through parts of both Baja California and Sonora—specifically through the municipalities of Mexicali and San Luis Río Colorado—and has a surface area of 5,923.16 km^2. The river borders the United States in the north, the Gulf of California in the south, the Altar desert in the east, and in the west the Cucapah Mountains, from which plains extend all the way to the river's mouth, where it forms a delta.

The aquifers located in the Mexicali Valley and Mesa Arenosa (Sandy Plateau) in San Luis Río Colorado, Sonora, are currently being exploited. The water extracted in this area is exported to Tijuana, where it is stored at the El Carrizo Dam in the municipality of Tecate.

Surface and Groundwater Quality

The greatest problem affecting the river is salinity (Singer 1998), given that an average of 10 tons of dissolved salts are conveyed downriver annually. These salts arrive at Imperial Dam, where the evaporative concentration reaches as much as 1 ton of salt per acre-foot. Average salinity is 750 parts per million (ppm) of soluble salts, or 0.75 grams of salt per liter of water. Thus, the Colorado River is a salty river, especially considering that other U.S. rivers such as the Columbia and the Mississippi contain 90 ppm and 200 ppm, respectively.

The Colorado River's salinity originates from three sources. The majority comes from the dissolution of salts in the sedimentary strata of tributary rivers flowing through the canyons of the Colorado Plateau. The second largest source is the leaching of salts from agricultural irrigation in the upper basin and the return flows that drain into the river (Singer 1998). The third source is intense water evaporation, particularly in the lower basin, where evapotranspiration occurs at a rate of 2,000 millimeters per year (mm/y). Salinity could become significantly worse if a greater volume of water is retained in the upper basin or if the average flow of the river decreases.

The first projects to control and convey surface water began early in the 20th century; salt content in 1902 was 400 ppm, by 1932 it had increased to 600 ppm, in 1963 it was 800 ppm, and by 1995 it had reached 1,050 ppm. This steady increase was due to the construction of controls, reservoirs, and dams. But above all, when the waters of the Colorado Basin were divided to increase the irrigable surface an intensive use of agrochemicals began. This caused the concentration and accumulation of salts, both in the groundwater and in the flows of the river (Sánchez and Mata 1997).

According to Mexico's Comisión Nacional del Agua (CNA, in English National Water Commission) (1995), the aquifer in the valley is overexploited—its deficit is 28,400 AF/y—and Mesa Arenosa is underexploited—its annual recharge and extraction are 121,600 AF/y and 82,700 AF/y, respectively. Of the surface water that enters the delta, Mexico's allocation, according to the 1944 Water Treaty, is 1.5 MAF/y. Of this allocation, 1.36 MAF/y is delivered directly to

Morelos Dam and 14,000 AF/y is diverted toward Lindero Sur (southern border). A 470 km major conveyance channel network begins at Morelos Dam, as does a 2,432 km minor conveyance network. Both are lined with hydraulic concrete.

The second-largest water reservoir for delta water is a shallow, non-isometric, free-type aquifer with a permeability of 0.1 square meters per second (m^2/s). The average annual extraction from the 658 wells in the Mexicali Valley aquifer and the 67 wells in Mesa Arenosa are detailed in Table 1. Obviously, the volumes provided by the two sources do not agree, making it clear the actual existing volumes and extraction rates need to be more accurately evaluated.

With regard to water quality in the aquifer, the increase in mineralization in the areas of Riito, Tulecheck, Aeropuerto, and Cerro Prieto, as well as the increase in salinity, could be attributable to geothermal water intrusion. The water quality of the aquifer located to the east, toward San Luis Río Colorado, is affected by a mixture of geothermal waters, seawater that evaporates and moves up into the valley through the fault system, as well as by seepage of Colorado River water. Instituto Mexicano de Tecnología de Agua (in English, Mexican Water Technology Institute) has been monitoring the aquifer for 30 years and concluded there is a 2% annual increase in mineralization, equivalent to 20 milligrams per liter (mg/L) (Sánchez and Mata 1997).

Table 1. Annual Average Extraction from the Mexicali Valley Aquifer and Mesa Arenosa

Source	Mexicali	Mesa Arenosa
Paredes 1992	750,640 AF/y	79,940 AF/y
GAS-CNA 1996	891,770 AF/y	121,605 AF/y

Notes: [1]Census rate 1900-1921: [2]Census rate 1930-1950
Sources: Unikel (1976), Arreola and Curtis (1993), INEGI population census 1990-2000, Sánchez and Mata 1997

Weather Characteristics

The Colorado River Delta belongs to the Sonoran Desert Geomorphic Province (Shreve and Wiggins 1964). According to the Köppen model as modified by García (1981), the entire province exhibits BW-type weather, which is dominated by very arid or very dry climates. This indicates a dry, warm desert with winter rains and very extreme temperatures. The annual median temperature is 22°C (72°F), annual precipitation averages 54.2 mm (2.13 inches), and the annual temperature variance is 17.7°C (63.9°F). These climatic characteristics do not, however, reflect the delta's extreme character. The data presented in Table 2 are from three climatological stations located throughout the valley and show peak maximum temperatures, peak minimum temperatures, and the minimum yearly precipitation that has occurred more than once since 1949 (Venegas 2000).

Geomorphology

The several subunits in this province are characterized by the plant species they host. Two are the Sonora-Arizona Plains, dominated by species with sarcocaule physiognomies, columnar cacti, and microphyllus leaves, and the lower Colorado River basin subunit (today the Imperial and Mexicali Valleys), with species adapted to living in lacustris and palustris environments, thus creating wetlands and

Table 2. Temperature, Precipitation, and Evaporation in Mexicali, the Delta, and Bataquez

Station	Maximum Temperature	Minimum Temperature	Annual Percipitation	Annual Evaporation
Mexicali	54.3°C (129.7°F)	-7°C (19.4°F)	54.2 mm (2.13 in)	
Delta	57°C (134.6°F)	-3°C (26.6°F)	36 mm (1.42 in)	2,160.3 mm
Bataquez	57°C (134.6°F)	-8.9°C (16°F)	33 mm (1.30 in)	2,369.0 mm

species that can withstand a high water table. This subunit is characterized by a nearly flat surface, making it a flood zone, with altitudes of more than 40 m where the river spills into the valley; the altitude progressively decreases until the Cucapah alluvial fan, at 14 m. Here the Cucapah mountains rise in a north-to-southeast direction and an altitude of 1,000 m.

The northeast portion of the lower Colorado River basin features the Mesa de Andrade dunes, which are located mostly within the United States. These dunes are created by winds and have maximum altitudes of approximately 100 m. Toward the northwest, the lower basin descends until it reaches sea-level, near the northwest limits of the city of Mexicali, almost at the international boundary with the United States. North from this point the altitude falls below sea level, reaching a low point of 87 m below sea level and forming the Salton Sea (see Figure 1). Toward the south the basin connects with the Gulf of California, which still exerts a great influence due to the activity of coastal processes. There are reports that during high tides, waves can reach between 3 m and 7 m. People who live near where the river flows into the sea call this process "el burro" (Tapia 2002). The coastal processes and waves created by high tides introduce sea sediments into the delta and mix saltwater with freshwater. Unfortunately, today no freshwater exists to dilute the seawater, which results in a significant increase in salinity (Luecke, et al. 1999).

On the western side, nearly against the Cucapah mountains, a volcanic structure called Cerro Prieto rises against a nearly flat landscape and has an altitude of 220 m. It is a Pleistocene volcano that last erupted 700,000 years ago. Today, it is inactive and around it lies the largest geothermal field in Mexico, which produces 750 megawatts of energy.

Soils

Soil studies are invariably conducted on a superficial layer, in most cases never greater than 2 m in depth. They allow the study of events such as geochemical, geomorphologic, and climatological processes. Characterizing the soils of the delta requires attention to the geological and geomorphologic details, as well as to the alternating glacial and interglacial periods. It also requires attention to intense Colorado River erosion, transportation, and sediment load accumulation events in the Salton Trough. This process lasted for at least 3 million years during the Plio-Pleistocene and involved the depositing of mostly fine sediments of sand, silts, and clays. These types of sediments from a more recent age make up the soils of alluvial origin. They exhibit great granulometric variance depending on the topographic characteristics of the areas traversed by the main flow of the river.

It is clear that the transformation of primary minerals into secondary materials, specifically in the case of clays, has not taken place in the delta. They were all transported by the river and continually renewed by new sediment contributions until they formed the flood plain as it is known today. This behavior defines the granulometrics of the sediments and their settlement, as well as the discordant layers of medium sands, followed by clays or silts, and so on successively, without a defined deposit sequence characteristic of a flood and sediment load accumulation area.

Based on this characterization, the soils of the delta are alluvial in origin with medium to fine textures. That is, the different strata are mostly layers of clays and silts, interspersed with fine sands. Due to these characteristics, the soils are deep, heavy, have poor drainage, high fertility, and are susceptible to salt intrusion. This type of soil would be considered an entisol.

Two works on soil in the Imperial Valley—Pierre, MacKenzie, and Zimmerman (1974) and Zimmerman (1981)—report the physical and chemical properties of the soils, their taxonomic classification, and define five series of soils. They also report the minimum area that can be mapped with individual characteristics for each, but conserve the characteristic order, which holds the highest rank under U.S. classification.

The series of soils determined in this study are essentially the same as those that have been mapped for the Mexicali Valley. The series are: Meloland, Imperial, Superstition, Holtville, and Gila. The main characteristics of these are alluvial origin, absence of diagnostic horizons, arid temperatures, medium to fine textures, poorly draining, and subject to alkalinity. According to Singer (1998), these soils are classified as *Typic Torriflivents*, *Xeric Torrifluvents*, and *Typic Torriorthents*; all belong to the order Entisols.

Water Stress

The Colorado River's delta and water are two of the most significant issues the border region faces today. Environmental groups, ecology groups, renowned scientists who have studied the region for many years, and several non-governmental organizations are extremely concerned about the current situation in the delta. Scientific and media reports, even as far back as the 1970s and 1980s, noted that the Colorado River was in danger of drying up (Luecke, et al. 1999). Today, the delta's area has decreased by 5% (Luecke, et al. 1999), due in part to the high concentration of mostly agricultural contaminants, the spread of which provides yet another sign that the surface size of the Colorado River Delta is shrinking.

The aforementioned risk notwithstanding, during the last 20 years there have been signs that this process could be reversed, given the recovery of nearly 55,000 hectares and due to the release of surpluses from U.S. reservoirs, agricultural wastewaters from both countries, and municipal waters from the city of Mexicali. Whether this trend can be maintained is uncertain. Consider also that these types of waters, coming as they do from municipal flows and irrigation districts in both valleys, contain large concentrations of salts and contaminants, but in spite of this have proven beneficial in the recovery of areas previously lost (Luecke, et al. 1999). However, above-average precipitation for most of the region, including the upper Colorado River basin, were due to a general condition caused by El Niño Southern Oscillation events, and there is no guarantee they will occur again.

This situation could also be to a new climatological phenomenon known as Pacific Decadal Oscillation. During this phenomenon, cycles of intense precipitation last for approximately 22 years, and then drought cycles occur of approximately the same duration. Under this new hypothesis of climatic behavior, the entire northwestern region of North America would experience drought until 2020 (Hare 2000).

Biotic Communities

A large number of scientific reports exist on the possible flora and fauna that the Colorado River Delta may have supported with its large volume of freshwater, vast amounts of nutrients through its sediment load, and the dynamics of intense waves in the Gulf of California. Some authors, such as Ezcurra, et al. (1999), estimate that between 200 and 400 species of vascular plants may have inhabited the delta.

When the first Spanish explorers arrived in 1540, they reported seeing jaguars, bears, deer, coyotes, and beavers (Luecke, et al. 1999). They also contacted the natives of this land, the Cucapah, who lived in the area surrounding the delta and had flourishing agriculture, including some variety of corn, beans, and squash. The Cucpah also collected a kind of grass called *trigo gentil* (gentle wheat) and may have used mesquite fruit, which they called *péchitas*, and hunted wild geese, ducks, and fish.

The environment experienced by the first explorers has changed dramatically from the environment seen by those who followed until the beginning of the 20th century due to the high degree of pollution, increase in salinity, and a nearly complete loss of freshwater. Completely stopping the contribution of sediments, and their corresponding nutrients, has caused a reduction of wetlands and the displacement of native riparian species such as poplars *(Populus sp.)* and willows *(Sallix sp.)*. Exotic species that are more aggressive and more resistant to high salinity, such as the Salt Cedar *(Tamarix ramosissima)*, which has a high evapotranspiration rate, took over. But regardless of the difficult situation in the Colorado River Delta, it remains the only safe habitat for a large number of migratory species because it is the only significant "fresh" water body among the wetlands of the Mexican Pacific.

Of the mastological fauna believed to have been present in the delta (Tapia 1997), thus far the river otter (possibly *Lutra* genus) and the mule deer *(Odocoileus hemionus)* have disappeared. However, Programa de Manejo de la Reserva de la Biósfera (PMRB, in English Biosphere Reserve Management Program) (INE-SEMARNAT 1995) cites the mule deer as currently inhabiting it. With regard the beaver *(Castor canadensis)*, Mellnik and Luévano (1995) report there are still populations in the delta, and although their status is not clear, the researchers affirm that if the level of freshwater is increased, the beaver population will increase as well. The lists of species presented by Instituto Nacional de Ecología-Secretaría de Medio Ambiente y Recursos Naturales (INE-SEMARNAT) (1995), Luecke, et al. (1999), and Tapia (1997) differ, thus this chapter will rely on Luecke, et al. (1999) preferentially because it is considered the most complete work on Colorado River Delta flora and fauna.

Per the Diario Oficial de la Federación (in English Mexican Federal Register) (1994), the endangered freshwater fish desert pupfish *(Cyprinodon macularius)*, and endangered bird, the Yuma clapper rail *(Rallus longirostris yumanensis)* are currently in the delta. The bobcat *(Lynx rufus)* and the totuava *(Cynoscion macdonaldi)*, which is the only endangered marine fish, as well as the Vaquita *(Phocoena sinus)*, the only endemic cetacean species in Mexico, whose distribution is restricted to the northernmost limit of the Gulf of California, are located nearby and are dependent indirectly upon flows from the Colorado River.

Luecke, et al. (1999) list 66 bird species, while the PMRB reports at least 80 species of land and water birds, both resident and migratory. The more frequent visitors include Osprey or fishing hawk *(Pandion haeliaetus)*, American white and brown pelicans *(Pelecanus erythrorhynchos* and *P. occidentalis)*, ring-billed gull *(Larus delawarensis)*, least tern *(Sterna antillarum)*, cormorants *(Phalacrocorax auritus)*, teal *(Anas crecca)*, Canada goose *(Branta candensis)*, and clapper rail *(Rallus longirostris)*.

In general, between 12 and 14 mammalian species are listed. These land mammals are representative of the Sonoran and San Bernardinian biotic provinces and include a broad diversity of rodents and species of interest for hunting, such as mule deer *(Odocoileus hemionus)*, foxes *(Urocyon cinereoargenteus, Vulpes macrotis)*, coyotes *(Canis latrans)*, and bobcats *(Lynx rufus)* (INE-SEMARNAT 1995). Because there is also disagreement on the number of species and their habitat; in this case, preference is given to the Programa de Manejo de la Reserva de la Biósfera.

Marine vegetation, made up mainly of algae and sea grasses, present in the delta includes *Distichlis palmeri*, a species endemic to the Gulf of California. The allophyllus vegetation is spread out over small areas along the coast, preferring closed-off, low-lying areas and marshes. They are represented by short plants, leafy succulents, and perennial grasses, such as Palmer's seaheath *(Frankenia palmeri)*, Mojave seablite *(Suaeda ramosissima)*, salty grass *(Distichlis palmeri)*, and alkali dropseed *(Sporobolus airoides)*. West of Adair Bay and east of the Santa Clara Marsh, marine vegetation is made up mostly of algae and sea grasses, most prominently *Distichlis palmeri*.

Conclusion

The Colorado River, which has fostered important industry, agriculture, urban areas, and tourism in both the United States and Mexico, faces a critical situation due to the grave hydrological imbalance to which it was subjected during the 20th century. It was then that the river was exploited without regard for its potential deterioration due to a reduction in its flow and the construction of large dams in the United States. However, if its flow had not been controlled and large dams had not been built to prevent its enormous floods, it would not have sparked such significant development, the growth of important cities, electricity generation, the irrigation of more than 500,000 hectares of crops in the Imperial and Mexicali Valleys alone, or provided water for residential use to several million residents in both countries. The question is, What would the region become if not for the water from this prodigious river? What actions could have been taken to avoid controlling it while still promoting development on the scale achieved by these projects?

Perhaps there are no right answers. Even with today's technological developments, uncertainties abound about how to face the new challenges posed by population growth in both countries and throughout the border region without an additional Colorado River or Río Grande to sustain the coming generations. The citizens of both countries, their scientists, administrators, and all who benefit must—beyond merely engaging in binational disputes over additional volume for the states that benefit from its wealth—work to preserve what little is left of what once was a great river and delta.

The solution lies not with rescuing or restoring habitats that existed before the 20th century. That would be largely impossible, given the great pressure being exerted by ever-increasing water and land demand in the border region of both countries. Between 1950 and 1995 the border region population in northern Mexico grew from 3.7 million to 15.2 million (Estrella, Canales, and Zavala 1999). Mexico's third largest border city, Mexicali, has an estimated population growth rate of 2.2% and will increase its population to 1.25 million residents by 2020. That nearly doubles its current population (Wright and Griffin 1998). This should motivate the development of new formulas and attitudes to make more efficient use of the scarce volumes expected to be available in the next two decades (Hare and Francis 1995; Mantua, et al. 1997).

References

Biondi, F., A. Gershunov, and D. Cayan. 2001. "North Pacific Decadal Climate Variability Since 1601." *American Meteorological Society* 14 (January): 5–10.

Comisión Nacional del Agua. 1995. "Plan Hidráulico Estatal, Gerencia Regional Península Baja California." http://www.cna.gob.mx.

Derrau, M. 1970. *Geomorfología*. Spain: Ariel.

Diario Oficial de la Federación. 1994. "Norma Oficial Mexicana, NOM-059-ECOL-1994." México, D.F.: Mexican federal government.

Elders, W., R. W. Rex, T. Meidav, P. T. Robinson, and S. Biehler. 1972. "Crustal Spreading in Southwestern California." *Science* 178 (4056): 15–24.

Estrella, V. G., A. Canales C., and M.E. Zavala. 1999. *Ciudades de la Frontera Norte: migración y fecundidad.* Mexicali, B.C.: UABC.

Ezcurra E., R. S. Felger, A. D. Russell, and M. Equihua. 1999. "Fresh Water Islands in a Desert Sand Sea: The Hydrology, Flora and Phytogeography of the Gran Desierto Oases of Northwestern Mexico." In *A Delta Once More: Restoring Riparian and Wetland Habitat in the Colorado River Delta*, D. F. Luecke,

J. Pitt, C. Congdon, E. Glenn, C. Valdés-Casillas, and M. Briggs, eds. Washington, D.C.: Environmental Defense Fund Publications.

García, E. 1981. *Modificaciones al sistema de clasificación climática de Köppen para adaptarlo a las condiciones de la República Mexicana.* México: UNAM.

Gastil, G., R. Phillip, and E. Allison. 1975. "Reconnaissance Geology of the State of Baja California." *The Geology Society of America Memoir* 140.

González, J. 1990. "La falla Imperial en el Valle de Mexicali (a 50 años del temblor Imperial magnitud Richter 7.0 del 18 de Mayo 1940)." *Comunicaciones Académicas* reporte técnico: crsrr 9003.

Hare, S. R., and R. C. Francis. 1995. "Climate Change and Salmon Production in the Northeast Pacific Ocean." Pages 357–372 in *Ocean Climate and Northern Fish Populations*, R. J. Beamish, ed. Ottawa: NRC Research Press.

Hare, S. 2000. "The Pacific Decadal Oscillation (PDO)." http://tao.atmos.washington.edu/pdo.

Instituto Nacional Ecología, Secretaría de Medio Ambiente y Recursos Naturales. 1995. "Areas naturales protegidas. Programa de la Reserva de la Biósfera Alto Golfo de California y Delta del Río Colorado." *Programa de Manejo* 1.

Lira, H. 1994. "Actividad sísmica registrada en el Valle de Mexicali, B.C., de 1973 a 1993." México: Comisión Federal de Electricidad.

Luecke, D. F., J. Pitt, C. Congdon, E. Glenn, C. Valdés-Casillas, and M. Briggs, eds. 1999. *A Delta Once More: Restoring Riparian and Wetland Habitat in the Colorado River Delta*, Washington, D.C.: Environmental Defense Fund Publications.

Mantua, N., S. R. Hare, Y. Zhang, J. M. Wallace, and R. C. Flores. 1997. "A Pacific Interdecadal Climate Oscillation with Impacts on Salmon Production." *Bulletin of the American Meteorological Society* 78: 106–1079.

Mellink, E., and J. Luévano. 1995. "Status del castor (Castor canadensis) en el Valle de Mexicali." Closing report, Programa Ambiental Frontera Norte-INE. Departamento de Ecología, Centro de Investigación Científica y de Educación Superior de Ensenada (CICESE).

Molina, R. 1991. "Sismología en el Valle de Mexicali." Travesía 27: 69–76.

Muffler, P., and B. Roe. 1968. "Composition and Mean Age of Detritus of the Colorado River Delta in the Salton Trough, Southeastern California." *Journal Sedimentary Petrology* 38(2): 384–399.

Ortega, J. 1995. "La evolución del ambiente: Paleoclimatología." Pages 9–19 in *IV Curso sobre Desertificación y Desarrollo Sustentable en América Latina y el Caribe*, August–September, Colegio de Posgraduados en Ciencias Agrícolas, Montecillo, México.

Pierre, E. R., A. J. MacKenzie, and R. P. Zimmerman. 1974. "Physical and Chemical Properties of Major Imperial Valley Soils." *Agricultural Research Service* April.

Puente, C. I. and A. de la Peña L. 1978. "Geología del campo geotérmico de Cerro Prieto." First Symposium on the Cerro Prieto Geothermal Field, Baja California, 20–22 September, San Diego, California.

Sánchez, O. 1990. "Crónica agrícola del Valle de Mexicali." Mexicali, B.C.: UABC.

Sánchez D., L. F., and I. Mata A. 1997. "Impacto de las actividades agrícolas en la calidad del agua subterránea del Distrito de Riego 014, Río Colorado, B.C." IMTA, Project TH-9614, Coordinación de Tecnología Hidrológica, May.

Shreve, F., and I. L. Wiggins. 1964. *Vegetation and Flora of Sonoran Desert*. Stanford, Calif.: Stanford University Press.

Singer, E. 1998. *Geology of the Imperial Valley*. In Keys to Soil Taxonomy, 8th Ed. Washington, D.C.: USDA-NRCS. http://soils.usda.gov/technical/classification/tax_keys/.

Sykes, G. 1937. "The Colorado Delta." *American Geographical Society* 19

Tapia, L. A. 2002. Personal communication with the author.

Tapia, L. A. 1997. Personal communication with the author.

Venegas, R. 2000. "El uso de la flora urbana en ciudades de clima árido seco extremoso." In *Ciudad, salud y medio ambiente*. Mexico: UAP-RNJU.

Wright, R., and E. Griffin. 1998. "The Imperial Valley-Mexicali Interface." http://geography.sdsu.edu/Research/Projects/Imperial/impweb.html.

Zimmerman, R. P. 1981. "Soil Survey of Imperial County, California, Imperial Valley Area." Washington, D.C.: USDA Soil Conservation Service.

II

Looking Beyond the Border: Environmental Consequences of the All-American Canal Project in Mexico and Potential Binational Solutions

Francisco Zamora Arroyo, Peter Culp, and Osvel Hinojosa Huerta

Introduction

The social and economic impacts associated with the proposed lining of the All-American Canal (AAC) are well described in other chapters. However, there are also significant environmental impacts that could be associated with the project. This chapter describes the potential impacts of the AAC lining project on the Mexican critical wetland habitat known as Andrade Mesa, as well as the legal implications of the AAC lining project that may be associated with these impacts under the National Environmental Policy Act (NEPA) and the Endangered Species Act (ESA). These implications were raised in a July 2005 lawsuit filed against the U.S. Bureau of Reclamation (BOR) by U.S. and Mexican plaintiffs.

The environmental and legal implications of the AAC lining project highlight the continuing need for discussions between the United States and Mexico on a number of important issues related to the Colorado River. In many ways, the AAC lining issue is a microcosm of larger transboundary concerns related to the environmental and socioeconomic consequences of Colorado River management. It deals with issues related to the impacts of reservoir operations on environmental values in the lower Colorado River and delta, the environmental and economic concerns with salinity and other water quality issues, and the ongoing conversation over the allocation of water in the 1944 Water Treaty that allocated Colorado River water between the United States and Mexico.

The dramatic impacts of the recent drought on reservoir storage throughout the Colorado River system have only increased the political pressures on the river resource. It is clear that, despite the relative abundance of the Colorado River in the past, the future of the river will be one of increasing pressure on water resources. Even assuming that the drought cycle that has characterized the first years of the new century will be of a short-term nature, the combination of progressively increased upper basin development with rapidly growing populations throughout the lower basin and Mexico will inevitably increase the pressure on the limited water resources of the Colorado.

In the context of growing pressure on water resources, the binational issues related to the Colorado River cannot effectively be addressed in a piecemeal fashion. Dealt with in isolation, many binational water issues appear to be effectively zero-sum, with the conversation focused on a discourse of entitlement—the rights of users in the United States versus the rights of users in Mexico, or the rights of humans versus the environment to water. This discourse inevitably frames controversies in a manner that encourages conflict, arms-length relationships, and inhibits open discussion.

This chapter suggests that the environmental concerns associated with the AAC lining project mandate that the United States and Mexico begin to approach the dispute over this and other issues in this broader context. Interests on both sides of the border can be served through a more collaborative approach to water management that addresses the concerns of both nations over the future of the Colorado River resource.

Background—The Lining Project

BOR constructed the AAC in the 1930s to deliver water from the Colorado River to users in California. The AAC has been in active use since the early 1940s and currently delivers an annual average of 3 million acre-feet (AF) of water from the Colorado River to the Imperial and Coachella Valleys. In 1988, a new law (Public Law 100-675) authorized BOR to line the AAC with concrete to save approximately 70,000 AF of water annually. This water is currently lost to seepage as the AAC traverses an area of sandy soil west of Yuma, Arizona, from Pilot Knob to a point known as Drop 3 (Figure 1). Funding for the project is expected to come from water agencies in California that have water rights to the Colorado River and will benefit from the conserved water.

The Final Environmental Impact Statement/Environmental Impact Report (FEIS/EIR) (BOR 1994) for the AAC lining project was issued in March 1994 pursuant to the requirements of NEPA. The corresponding Record of Decision (ROD) was signed on July 29, 1994. The FEIS/EIR considered five alternatives for the AAC project:

1. A new, concrete-lined canal parallel to the existing AAC between a point 1.6 kilometers (km) west of Pilot Knob and Drop 3 (approximately 37 km in length).
2. In-place canal lining between Pilot Knob and Drop 3 (approximately 39.6 km in length).
3. In-place lining from Pilot Knob to Drop 4 (48.1 km in length).
4. Groundwater pumping along the sides of the AAC to recover canal seepage (as an alternative to direct seepage control via canal lining).
5. A no-action alternative under which no seepage controls would be implemented (consideration of a no-action alternative is required by NEPA).

The FEIS/EIR selected the parallel canal as the so-called preferred alternative due to its comparatively lower costs. The total construction costs were estimated to be approximately $86 million in 1994, with annual operation and maintenance costs increasing by

Figure 1. All-American Canal and Proposed Lining Project.

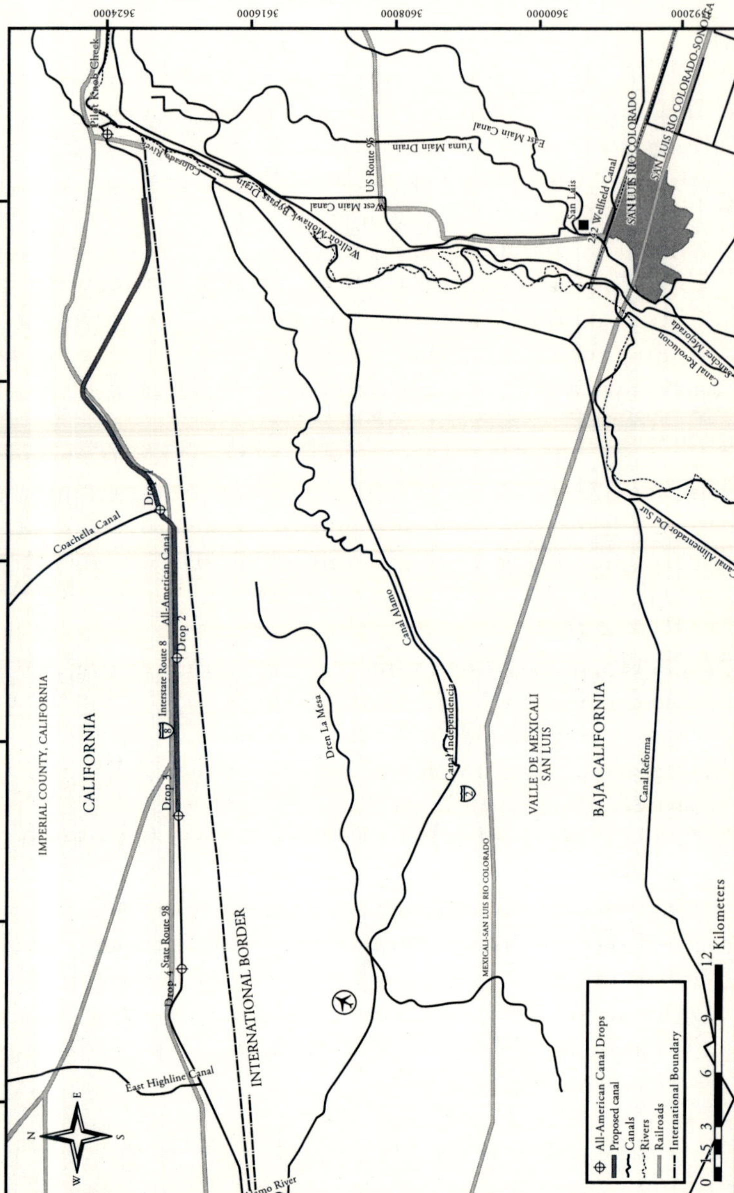

approximately $14,000 to $279,000 annually. The alternative was estimated to conserve 67,700 AF annually, at a calculated cost per acre-foot conserved of $109 (based on recovery of total construction costs and 8% interest accrued during construction over a 50-year period, divided by the amount of water expected to be conserved annually).

The 1994 FEIS/EIR and ROD underwent a five-year required reexamination by BOR in May 1999. BOR found that "the FEIS/EIR and ROD for the All-American Canal lining project continues to meet the requirements of the National Environmental Policy Act and the California Environmental Quality Act and that it should be valid until completion of proposed construction in 2006." BOR also found there were no changes in the purpose of the project or in the preferred alternative. A new periodic reexamination was required by 2004. Although some occasional references to this document have been encountered in other materials associated with the AAC project, the authors were unable to locate a copy despite repeated inquires with BOR staff.

Environmental Impacts in the United States

The FEIS/EIR considered potential environmental impacts to four wetland areas, all of which were within the territorial boundaries of the United States and located within or in close proximity to the AAC (BOR 1994):

1. A 575.4 hectare wetland complex between Drops 3 and 4. Of the total hectares, 305.5 (53%) consist primarily of salt cedar, 44.9 (9%) are marsh areas, 94.3 (16%) consist primarily of arroweed, 101.6 (18%) are mesquite, 15.8 (3%) are riparian vegetation of cottonwood and willow, and approximately two hectares are open water.
2. A 40.5-hectare, seepage-induced wetland along both sides of the AAC between Drops 2 and 3. Most of this wetland consists of arroweed (50%) and salt cedar (30%), with only 20% of the area consisting of mesquite. The area also contains approximately two-fifths of a hectare of marsh.

3. Some 1.9 hectares of open water associated with recapture ditches along the sides of the AAC.
4. Continuous stands of common reeds, between less than 1 meter (m) and 4.6 m in thickness, which grow along the AAC for most of the distance from Pilot Knob to Drop 4 (approximately 12.1 hectares total).

Proposed Wetlands Mitigation in the United States

Although the Drop 4 alternative (not selected) would have saved an additional 1,000 AF by lining an extra 8.5 km of the canal, the mitigation costs associated with this alternative were considerably higher due to the projected impacts to the 575.5 hectares of wetlands located between Drop 3 and Drop 4 and the requirements of California's "no net loss of wetlands" policy. Although the projected wetland mitigation costs were not precisely broken out in the FEIS/EIR, a rough estimate of the wetland mitigation costs can be derived by comparing the differences between the two alternatives and the projected overall costs of each alternative. Considering the difference in projected costs between the Drop 3 and Drop 4 alternatives, and subtracting the additional construction costs associated with the Drop 4 alternative, the mitigation costs associated with the 575.5 hectares of wetlands between Drop 3 and Drop 4 would have been at least $10 million.

The preferred alternative avoided any impact to the 575.5 hectare wetland between Drops 3 and 4; as such, only minimal wetlands mitigation was deemed to be required. Wetlands mitigation under the preferred alternative consisted of the establishment of 17.4 hectares of honey mesquite and cottonwood/willow riparian habitat (at a density of approximately 50 trees per hectare) and the creation of two fifths of an additional hectare of marsh in the existing Drop 3/Drop 4 wetland complex. Mitigation sites were to be selected based on physical and biological suitability criteria, avoidance of impacts to existing vegetation, and maximization of value to wildlife species of special regulatory concern (including Yuma clapper rail and California black rail) (BOR 1994).

Mitigation for vegetation along the canal that was projected to be lost as a result of the canal lining consisted of providing replacement habitat for special status species. Fish mitigation under the alternative consisted of the construction of artificial reef areas (discarded tire reefs) in the new lined canal, built to U.S. Fish and Wildlife (FWS) specifications.

To compensate for what were presumed to be negligible but potentially cumulative impacts to wetlands along the Colorado River, the preferred alternative also implemented on-river mitigation. Specifically it assisted BOR with funding for new lower Colorado River backwater restoration or enlargement projects. The Bureau of Reclamation (1994) identified the following mitigation costs (in 1994 dollars) associated with the preferred alternative:

Wetlands-related mitigation	
Lower Colorado River backwater restoration	$100,000
Wetlands mitigation (Drop 3/Drop 4 and replacement habitat)	$150,000
Tire reef placement and construction	$250,000
Wetlands mitigation subtotal	$500,000
Other mitigation	
Replacement habitat for flat-tailed horned lizard/sand dunes	$751,500
Escape ridges for large mammals/ drowning prevention	incl. in canal cost
Archaeological surveys	$100,000
Soil stockpiling/recontouring	$60,000
Recreation plan	$5,000
Total mitigation costs (including wetlands)	$1,416,500
Estimated operations and maintenance costs	
For all proposed mitigation	$46,000/yr

Likely Environmental Impacts in Mexico

Impacts to wetlands in Mexico were not considered in the FEIS/EIR. Although the FEIS/EIR notes that the lining of the canal would cause impacts to groundwater in Mexico (exacerbating projected groundwater declines due to over-pumping groundwater for agriculture in the Mexicali Valley), the report does not identify any specific environmental or ecological impacts associated with the loss of seepage. The FEIS/EIR estimates the total contribution of AAC seepage between Pilot Knob and Drop 4 at approximately 91,600 AF per year (AF/y), which has created a large groundwater mound underneath the canal. Based on rough studies of the area hydrology, the FEIS/EIR estimates that approximately 90% of this seepage drains south toward the international border and that the other 10% flows north toward the East Mesa area (there are only negligible evapotranspiration losses).

Overall, the FEIS/EIR estimates that the total canal seepage accounts for approximately 12% of groundwater recharge to the undefined groundwater aquifer in the area. This aquifer, according to the FEIS/EIR, extends east under the Colorado River as far as Yuma, west at least as far as Drop 3, north under the East Mesa, and south across the international boundary into Mexico. Implementation of the preferred alternative is expected to result in localized groundwater declines in this aquifer (due to the decline in the existing groundwater mound under the AAC) to pre-AAC levels (from 0 m to 3 m below the canal at present to 12.2 m to 24.4 m below the canal) (BOR 1994). The FEIS/EIR estimates that on a larger scale, implementation of the preferred alternative would result in water table declines of less than 1 m to 9 m over a 112.7 km^2 area in the northeastern portion of the Mexicali Valley over a period of 50 years, assuming no change in the levels of groundwater pumping in the Mexicali Valley.

Assessment of Environmental Impacts in Mexico

Although not recognized in the FEIS/EIR, the lowering water table is likely to have an extreme impact on several significant wetland and terrestrial areas in Mexico. The Andrade Mesa wetlands were discovered only recently and initially documented in 2002. A rapid assessment study conducted in 2005 by ProNatura Noroeste-Sonora and the Sonoran Institute documented in more detail the extent and type of habitat in the Andrade Mesa wetlands using a combination of remote sensing techniques and field surveys (Zamora Arroyo, et al. 2005a; Zamora Arroyo, et al. 2005b). In addition, in 2004

ProNatura conducted a rapid assessment of bird species to better understand the ecological significance of these wetlands (Hinojosa Huerta, et al. 2004). The description that follows is based on these reports.

To facilitate comparisons between the impacts in the United States and Mexico, the classification used here for habitat is consistent with the classification used by the FEIS/EIR in the United States. This classification assumes wetlands habitat includes marsh and riparian areas. Marshes consist of open water, cattail, and other emergent vegetation that require water-saturated soils, while riparian areas include arroweed, salt cedar, mesquite, willows, and cottonwoods. Terrestrial habitat was defined to include desert shrubs.

The Andrade Mesa wetland area consists of approximately 1,921 hectares of wetland habitat and 1,453.2 hectares of terrestrial habitat; the entire area totals 3,374.2 hectares. Out of the 1,921 hectares of wetlands, 212.5 consist of marsh, 66.3 are arroweed and salt cedar, and the remaining 1,641 are riparian habitat with at least 10% coverage of mesquite (Table 1). Wetland areas are distributed in two major wetlands complexes, the Ejido Irapuato and the Ejido Nezahualcóyotl (Figure 2).

To determine the likely environmental impacts on the wetland and terrestrial habitats associated with the proposed lining of the AAC, the authors defined a likely impact zone based on existing information on seepage flows. Although the study did not include an analysis to determine the relationship between seepage from the AAC and the Andrade Mesa, the physical proximity of the wetlands to the AAC, the relative elevation of the AAC in relation to the wetlands, anecdotal evidence derived from the construction of the Mesa Drain in the 1960s strongly suggests that the AAC is a major source of water for these wetlands. This drain intercepts seepage from the AAC to prevent water saturation of farmland located south of the wetlands. Other studies have reached similar conclusions. According to Díaz Cabrera (2001), the underground water from Pilot Knob to Drop 1 flows to the south into the Colorado River. According to Cortez Lara and García Acevedo (2000), from Drop 1 to Drop 3 groundwater flows in a south-southwest direction. Based on 2004 piezometric observation along a north-south profile, the Comisión Nacional del Agua (CNA, in English National Water Commission)

Table 1. Type and Extent of Habitat in Mexico Likely to be Impacted by Lining the All-American Canal

Habitat	Total Hectares	Impacted Ares (hectares)	Percent Area Likely to be Lost
Wetland (marsh, arroweed, salt cedar, mesquite-creosote)	1,921.2	1,881.2	98
Desert shrub (diverse species <10% mesquite)	1,454.0	388.8	27
Total habitat	3,375.2	2,270.0	67
Vegetation/ Cover Type	**Total Hectares**	**Impacted Ares (hectares)**	**Percent Area Likely to be Lost**
Open water	23.7	23.7	100
Salt grass	52.6	52.6	100
Cattail	136.3	136.3	93
Arroweed and salt cedar	66.6	66.6	100
Desert shrub (10% to 30% cover)	864.1	220.5	26
Desert shrub (30% to 70% cover)	589.9	168.3	29
Mesquite and creosote (10% to 30% cover)	599.8	599.8	100
Mesquite and creosote (30% to 70% cover)	1,042.3	1,011.6	97
Total, all classes	3,375.3	2,279.4	67

Figure 2. Type of Habitat in the Andrade Mesa and the Identified Impact Zone Associated with Lining the All-American Canal

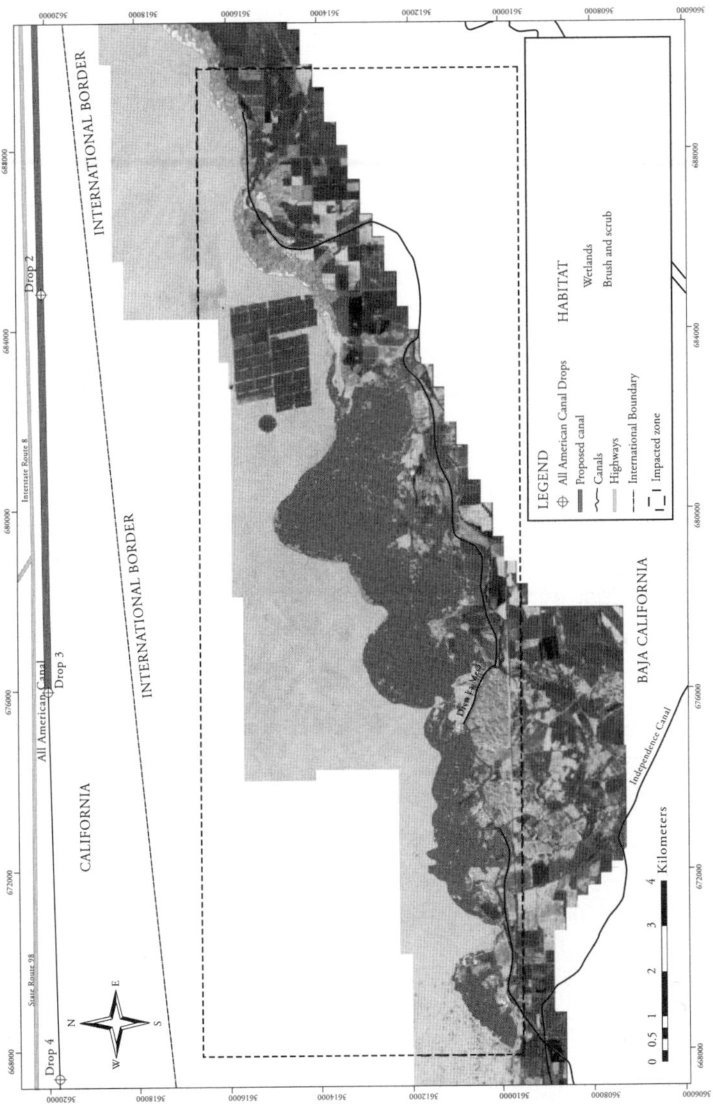

Source: Francisco Zamora

(CNA 2005) estimated that the hydraulic difference between the AAC and the Mesa Drain is approximately 12 m and suggested that the seepage from the AAC is a main source of water for the Andrade Mesa wetlands before water is captured by the Mesa Drain. With the construction of the new lined canal, the groundwater mound that has formed beneath the canal will decline within the next 10 years to 15 years, which will impact the quantity and quality of the aquifer as well as eliminate the source of water for the wetlands (CNA 2005). Given the evidence that a connection may exist and the significance of the potential impacts, the precise nature of the relationship between the AAC and the wetlands should be established by a supplemental environmental impact statement (EIS).

The authors identified the likely impact zone in Mexico as the area south of AAC between Drop 1 and Drop 3 (Figure 2). As a result, the proposed AAC lining project is considered likely to impact 1,881 (98%) of the wetland habitat and 388.5 (27%) of the terrestrial habitat in the Andrade Mesa (see Table 1). It is estimated that the 203.6 hectares of marsh wetlands, with open water, cattail, and salt grass areas (Figure 3), is likely to be the first habitat eliminated because these elements depend on open water or saturated soils conditions. Riparian vegetation would be maintained in the short term by shallow water tables, but in the long term it would disappear as local water tables decline.

Ecological Value of the Andrade Mesa Wetlands

ProNatura has systematically monitored bird populations in the Andrade Mesa wetlands with standard protocols. A total of 101 species of resident and migratory birds have been documented (Table 2). Among these, the species have federal protection status in the United States (all are endangered), 10 species are protected in the state of California (three are endangered, two are threatened, and five have special protection), and six species are under protection in Mexico (one is endangered, one is threatened, and four have special protection) (Hinojosa Huerta, et al. 2004) (Table 3). Among the listed species present in the wetland area are the Yuma clapper rail (Rallus longirostris yumanensis), which is endemic to the lower Colorado River basin and delta. The clapper rail is protected in

Figure 3. Marsh Wetland in the Ejido Irapuato Wetland Complex

Mexico as a threatened species and is listed as an endangered species in the United States. The population of Yuma clapper rail in the Andrade Mesa wetlands is estimated at approximately 170—the second largest population in Mexico after the Ciénega de Santa Clara (Hinojosa Huerta, et al. 2004). Another species present in the wetland area is the black rail (Laterallus jamaicensis coturniculus), which is listed as an endangered species in Mexico and California and is a candidate species under consideration for listing in the United States.

Table 2. Bird Groups Present in the Andrade Mesa Wetlands

Group	Number of Species	Total Counts	Percentage of Counts
Aquatic birds	43	2,138	50.50
Terrestrial birds dependent on the wetlands	9	187	4.41
Desert resident birds	21	1,680	39.77
Raptors	8	24	0.56
Migratory neotropical land birds	17	198	4.64
Exotic birds	2	7	0.12
Total	100	4,234	100.0

Table 3. Bird Species of Conservation Interest in the Andrade Mesa Wetlands

Species	Mexico	U.S.	California	Andrade
Rallus longirostris yumanensis	T	E	T	Rr, CO
Laterallus jamaicensis coturniculus	E	NL	T	Rr, RA
Passerculus sandwichensis rostratus	SP	NL	SP	Rr, PC
Ixobrychus exilis	NL	NL	SP	Rr, CO
Sterna nilotica	NL	NL	SP	Vn, PC
Rallus limicola	SP	NL	NL	Rr, CO
*Athene cunicularia**	NL	NL	NL	Rr, PC
Falco peregrinus	SP	E	E	Vn, RA
Empidonax traillii	NL	E	E	Vn, PC
Accipiter striatus	SP	NL	SP	Vn, PC
Lanius ludovicianus	NL	NL	SP	Rr, CO
Melanerpes uropygialis	NL	NL	SP	Rr, RA

Notes: Protection status in Mexico (MX), the United States (US), and California (Cal): T—threatened, E—Endangered, SP—Special protection, NL—Not listed. Use: Rr—Reproductive resident, Vn—Not reproductive visitor, CO—Common, PC—Present but uncommon, RA—Rare. *denotes a priority species for the North America Commission for Environmental Cooperation.

The presence of the Yuma clapper rail and the black rail populations in this area should be considered extremely significant because the lining project is likely to eliminate the second largest population of both subspecies in Mexico. The likely impacts to the wetlands will also affect species of migratory terrestrial and water birds by eliminating another important stopover site along a major migratory route on the Pacific Flyway. During the surveys conducted in the area, a large number of water birds were present, particularly waterfowl (14 species) and shorebirds (10 species). An estimated 10,000 birds (at minimum) use these wetlands during the annual migration (Hinojosa Huerta, et al. 2004). The reduction of terrestrial habitat would also likely decrease the population density of resident desert birds. As a result of these findings, these wetlands have recently been identified by a team of scientists as a conservation priority in the Colorado River Delta (Zamora Arroyo, et al. 2005c).

Due to their significant ecological value to the Colorado River Delta ecosystem, the Andrade Mesa wetlands would ideally be protected and maintained in its current condition. However, if the AAC lining is implemented, these wetlands will likely disappear absent a significant engineering project to artificially recreate the seepage flow into the area.

The existing FEIS/EIR suggests a mitigation approach for the other wetlands anticipated to be affected by the AAC lining project based on the guidelines by Anderson and Ohmart (1984), which assign a given ecological value depending on the type of habitat. Table 4 summarizes the value assigned to each habitat type. For example, it was determined that marsh wetland should be mitigated at a 1:1 ratio, and therefore there is no ecological value assigned to this habitat type. On the other hand, the ecological value for riparian habitat (salt cedar-arroweed) is 5.37, for cottonwood-willow riparian forest is 21, for mesquite is 10, and for desert shrub is 5, with their equivalences with other habitat types shown in Table 4.

Table 4. Ecological Value by Habitat Type and Their Equivalence with Other Habitat Types Used for Identifying Mitigation Requirements and Alternatives in the FEIS/EIR

Habitat Type	Hectares	Ecological Value	Equivalence Riparian (+)	Equivalence Mesquite
Marsh	203.60	NA	NA	NA
Riparian (-)	66.40	5.37	3.91	1.86
Mesquite	1,611.10	10.00	2.10	1.00
Desert shrub	157.40	5.00	4.20	2.00

Note: Riparian (-) refers to salt cedar-arroweed areas, while riparian (+) refers to cottonwood-willow forests.

Based on their habitat value and their equivalences, the mitigation for the Andrade Mesa wetlands associated with the FEIS/EIR would consist of the following (see Table 5):

- Creating 233.6 hectares of marsh wetland to replace impacted marsh habitat at a 1:1 ratio
- Creating 371.5 hectares of mesquite forest and 699.7 hectares of riparian forest of cottonwood and willow to mitigate for 1,611.1 hectares of impacted mesquite, riparian (-), and desert shrub habitat

An initial estimate of potential mitigation costs indicates this or similar mitigation would be $10 million or higher. Recommended areas for mitigation actions are described later.

Table 5. Ecological Value and Mitigation for the Different Habitat Types to be Impacted by Lining the All-American Canal

Habitat Type	Existing Habitat to be Impacted (hectares)	Ecological Value by Habitat Type	Ecological Value for Total Existing Habitat Type	Recommended Mitigation (hectares)	Ecological Value of Mitigation
Marsh	203.6	NA	NA	203.6	NA
Riparian (arrowed and salt cedar	66.4	5.37	880	0	0
Riparian (cotton-woods and willow		21		699.7	36,309
Mesquite	1,611.1	10	39,810	371.5	9,180
Desert shrub	388.5	5	4,800	0	
Total	2,269	41.4	45,490	1,274.8	45,490.0

Note: NA = does not apply because it is replaced 1:1

Legal Implications of the National Environmental Protection Act, Endangered Species Act, and the Migratory Bird Treaty Act

The FEIS/EIR indicates that the United States consulted Mexico about lining the AAC pursuant to the requirements of International Boundary and Water Commission (IBWC) Minute 242, Point 6. It appears, however, that the existence of the Andrade Mesa wetlands—and the suspected hydrological connection between the AAC canal seepage and those wetlands—was not discovered or disclosed during the course of this consultation. As noted above, neither the FEIS/EIR or the subsequent recertification actions have considered any potential impacts to the Andrade Mesa wetlands or even so much as recognized their existence.

The Council on Environmental Quality (CEQ) regulations, which were adopted pursuant to NEPA, define two circumstances that require the preparation of supplements to draft or final environmental impact statements:

- The agency makes substantial changes in the proposed action that are relevant to environmental concerns
- There are significant new circumstances or information relevant to environmental concerns and bearing on the proposed action or its impacts[1]

Factors to be considered in determining whether to prepare a supplemental EIS based on new information include, among others, the environmental significance of the new information, the degree of care with which the agency considered the information and its impact, the degree to which the agency supported its decision with a statement of explanation or additional data, and whether the action would affect threatened or endangered species.[2] Generally, where new information standing alone would constitute a "significant" impact requiring the preparation of an EIS, a supplement is required.[3]

Given the scale of the wetlands complex at Andrade Mesa, there is no question that the existence of the wetlands—and the potential for a hydrological connection with the AAC—constitutes "new circumstances or information" within the meaning of the CEQ regulations. BOR has already recognized the "significance" of the U.S. wetland complexes associated with the AAC in the FEIS (wetland areas that are considerably smaller than the Andrade Mesa complex). As such, a finding that impacts to the Andrade Mesa wetlands in Mexico are not significant would be unsupportable, particularly in light of the fact that BOR has already reissued the draft EIS for lining the Coachella Canal. The latter project, which is integrally connected to the AAC lining project as a conditional element of California's Quantification Settlement Agreement (QSA), will have relatively minor environmental impacts but nevertheless was determined to merit reconsideration by virtue of its inclusion in the QSA.[4] Even if these considerations could be overlooked, however, the fact that impacts to the Andrade Mesa wetlands may well harm threatened and endangered species will unequivocally require a supplement under the CEQ regulations.[5]

As such, the preparation of a supplemental EIS seems to be clearly required—at least insofar as NEPA would apply to an action in the United States that causes environmental impacts in Mexico. Although NEPA's applicability to transboundary environmental impacts was a subject of some controversy during its early history (owing to a general presumption in the judicial interpretation of American law against extraterritorial application in the absence of an affirmatively contrary Congressional intent),[6] recent interpretations of NEPA make it abundantly clear that NEPA requires consideration of transboundary impacts. In 1997, CEQ adopted a formal memorandum that interpreted NEPA to require federal agencies to analyze transboundary environmental impacts based on the express provisions of the law and judicial precedent. CEQ found that NEPA "directs federal agencies to analyze the effects of proposed actions to the extent they are reasonably foreseeable consequences of the proposed action, regardless of where those impacts might occur... [including] indirect effects ... regardless of geographic boundaries within the United States."[7]

Although the United States has more recently taken the position that NEPA should be limited to impacts within its borders, the courts have found otherwise. The Bush Administration argued that NEPA does not apply outside U.S. territory in cases involving the impacts of U.S. Navy sonar testing on marine mammals, the impacts of acoustical research by the National Science Foundation in the Gulf of California, and a case involving a proposed transfer of water from the Missouri River basin to the Hudson Bay river basin in North Dakota that would have impacted Canadian fisheries. In each case, the courts upheld the applicability of NEPA to the proposed federal action.[8] These decisions are consistent with precedent finding NEPA applicability in transboundary or extraterritorial cases. In several cases, these decisions granted standing to foreign plaintiffs to challenge the adequacy of NEPA analyses where proposed actions would cause impacts in foreign territories.[9] This interpretation is consistent with a generally accepted principle of international law—that no nation may undertake actions on its own territory that will harm the territory of another state.[10]

As such, it would appear that NEPA requires consideration of the potential environmental impacts in Mexico associated with AAC lining. This conclusion is consistent with the recent court decision in *Defenders of Wildlife v. Norton*,[11] which considered whether BOR was required to engage in a formal consultation under ESA with regard to the impacts of U.S. dam operations in the Colorado River Delta. Indeed, in that case the court reaffirmed that the presumption against the extraterritorial application of U.S. statutes generally, and ESA in particular, is "inapplicable ... to federal agency actions within the United States that have extraterritorial effects."[12] Although the court ruled in *Norton* that no consultation was required, the ruling was based on a finding that BOR lacked discretion to operate the dam system in a manner that would deliver water to Mexico for the benefit of the delta other than as provided in the 1944 Water Treaty. Nor could it guarantee that additional water delivered to Mexico would actually reach the delta. As such, the environmental impacts associated with U.S. dam operations in Mexico were not within BOR's discretionary control and were not subject to consultation, because this consultation would not serve ESA's purpose (because BOR lacked the discretion to modify its actions in order to mitigate for the identified impacts).[13]

Unlike the *Norton* case, however, the manner and extent of the AAC lining project—and the environmental impacts on the Andrade Mesa wetlands associated with this project—are largely within BOR's discretionary control. In addition, unlike ESA, which requires the agency to take affirmative action to mitigate for the impacts, NEPA requires only that these impacts be considered. As such, the purpose of the law is served by the agency's consideration of the impacts regardless of whether the agency can effectively prevent or mitigate those impacts.

Even more significantly, the act authorizing the canal lining also authorizes the Secretary of the Interior to "implement measures for the replacement of incidental fish and wildlife values adjacent to the canals foregone as a result of the lining of the canal or mitigation of resulting impacts on fish and wildlife resources from construction of a new canal, or a portion thereof."[14] This fact is particularly significant because it heightens the potential that an assessment of environmental impacts in Mexico pursuant to the requirements of NEPA could raise ESA compliance requirements as well, given that impacts on endangered and threatened species may well be identified in association with the impacts on the Andrade Mesa wetlands. Given the increased discretion associated with BOR's activities on the AAC, and given the express statutory authorization to engage in mitigation activities as a part of the AAC lining project, BOR consultation with FWS could be required.

It should also be noted that of the species present in the wetlands, approximately 91 are protected under the Migratory Bird Treaty Act (MBTA). Since BOR recertified the FEIS/EIR in 1999, a federal appeals court has ruled that federal agencies are subject to the MBTA's take prohibitions (*Humane Society v. Glickman*, 217 F.3d 882). That federal agencies are subject to the permit requirements of the FWS's existing regulations is now reflected in the FWS MBTA manual. Also, after 1999, U.S. President Bill Clinton signed an executive order that directs federal agencies to take certain actions to implement MBTA, including the development of a Memorandum of Understanding (MOU) with FWS to promote the conservation of migratory bird populations and minimize take (harm, kill, or destroy habitat) of protected birds. BOR is still in the process of drafting its MOU.

On July 19, 2005, Mexico-based Consejo de Desarrollo Económico de Mexicali, A.C. (CDEM) and California-based Citizens United for Resources and the Environment (CURE) filed a complaint in Nevada District Court against BOR over the AAC lining project, alleging a series of claims related to violation of water rights as well as violations of NEPA, ESA, and MBTA. Should this lawsuit prove successful, it could require the preparation of a supplemental EIS for the AAC lining project, as well as compliance with the requirements of ESA and MBTA.

The lawsuit may ultimately prove disruptive to Colorado River-related operations in California, particularly with regard to the fulfillment of settlement obligations to the San Luis Rey tribe and the implementation of water transfers envisioned by the QSA. This, in turn, could put the QSA at risk and potentially unravel this critical settlement, with destabilizing implications for lower Colorado River operations generally.

The Broader View: Opportunities for Binational Cooperation

The binational environmental issues surrounding the AAC lining project, at least as it is currently constructed, may well prove intractable. At its core, the dispute between U.S. and Mexican interests over the AAC lining project is currently framed in a zero-sum context—whether the water that currently leaks from the AAC will continue to be available to support groundwater tables in Mexico and the Andrade Mesa wetland complex, or whether this water will be conserved and thus made available to water users in California.

In this sense, the AAC lining project can be viewed as a microcosm of the larger suite of binational environmental issues associated with the Colorado River and its delta. A recently-released report, "Conservation Priorities in the Colorado River Delta, Mexico and the United States," identifies a series of environmental values associated with the Colorado River Delta. As detailed in the report, available at http://www.sonoran.org, scientists have identified approximately 15 priority conservation sites throughout the Colorado River Delta region that have particular significance as a part of the larger delta ecosystem.

Aside from their environmental importance, all these sites also are currently or imminently threatened by development, water efficiency improvements, specific water conservation or water use projects (such as the AAC lining project), or other changes in the prevailing regime for the distribution and use of water. Throughout the delta ecosystem, environmental values are almost always dependent to a greater or lesser degree on "waste" flows of water—water that has either become too salty for economical agricultural use or water otherwise unnecessary or lost to the system through canal leakage, excess irrigation, drainage, or other sources.

Many of these resources are directly or indirectly dependent on current system "inefficiencies" in the United States that result in water deliveries to Mexico beyond those that are legally required—

the AAC canal leakage, bypass flows through the Main Outlet Drain Extension (MODE) canal (necessary to meet water quality limits), occasional flood events that cannot be captured in the U.S. reservoir system, and overdeliveries as a result of cancelled or changed water orders. However, under the allocation framework of the 1944 Water Treaty, the continued existence of water from these sources is by no means guaranteed. The framework established by the treaty is ultimately quite rigid—giving Mexico the right to no more than 1.5 million AF of Colorado River water each year—an allocation that is essentially fully dedicated to agricultural and municipal use in Mexico.

While the combination of bypass flows, leakage, floods, and overdeliveries have historically delivered water well in excess of this amount for the benefit of the delta ecosystem, pressure from U.S. users to eliminate these inefficiencies continues to grow, particularly in the context of the current drought situation where lower basin users now face the potential for shortages that could be mitigated or prevented by limited excess deliveries to Mexico. Although U.S. environmental laws may provide the basis to retard or limit the scope of proposed efficiency improvements for the benefit of environmental resources in the delta, the fact of the matter is that so long as these issues continue to be framed in a zero-sum context, these environmental resources will be under continuing threat.

Viewed in a broader context, however, the United States and Mexico are facing a series of challenges related to water management in the border region that may not, in the aggregate, be zero-sum. At the root of many of the socioeconomic and environmental issues in the border region is the need for appropriate quantities and qualities of water to support human and environmental uses. There are interests on both sides of the border to be served through a more collaborative approach to water management that could identify solutions that avoid zero-sum resolutions. Such an approach could help resolve not just the issues related to the AAC lining project, but the important environmental and socioeconomic concerns associated with other Colorado River resources as well.

Some of these options would not necessarily require any significant changes to the current framework of the 1944 treaty. Over the years, there have been a number of proposals suggesting means by which the United States and Mexico might cooperate to improve the quality and quantity of water to support human and environmental uses within the limits of the water allocation regime provided by the treaty. Many of these proposals have focused on opportunities to improve the efficiency and quality of agricultural water delivery and use in the border region. For example, in 1991, BOR and CNA released a joint proposal entitled "International Cooperative Water Conservation and Irrigation Efficiency Improvement Program between the Republic of Mexico and the United States of America," which was championed by Dennis Underwood, the now-deceased CEO of the Metropolitan Water District of Southern California.

In this proposal, BOR and CNA identified a series of common national goals related to agricultural water conservation and achieving water system efficiencies to meet the future water needs of both countries. The proposal noted that a variety of water conservation and efficiency improvements were underway in the Imperial Valley, funded by a creative non-federal revenue source that allowed municipal and industrial users in California to invest in efficiency improvements that might otherwise be beyond the financial means of farmers in the Imperial Valley. The proposal also noted that the Mexicali Valley has cropping patterns and irrigation methods similar to those used in the Imperial Valley, and that the growing need for municipal and industrial water supplies in the Mexicali Valley and the Tijuana area is similar to the experience of Southern California. However, unlike efforts in the United States, current needs for expanded municipal and industrial supply and wastewater treatment facilities had significantly diminished Mexico's ability to finance similar improvements in Mexicali Valley agriculture.

To improve the financing capability for these improvements, the proposal suggested a cooperative program that would provide short-term water supply benefits to the United States and both short- and long-term water supply benefits to Mexico. Under this program, a revenue stream for efficiency enhancements would be provided by public water agencies in the United States that need temporary use of conserved water. During the period when irrigation works and efficiency improvements were being constructed and implemented for a given block of land in the Mexicali Valley, irrigation would be suspended and there would be a corresponding reduction in water use. In exchange for the temporary use of this unused water, U.S. agencies would fund the construction of works and efficiency improvements while contributing funds to offset short-term farm income reductions among Mexican water users.

These enhancements would be phased over a number of years, providing a continuing revenue stream for improvements in Mexico without disrupting local economies, while making a temporary pool of water available to U.S. users. Over the long term, this program would result in the rehabilitation and modernization of the agricultural distribution system in the Mexicali Valley; provide both countries with access to an international pool of technical expertise in water resources management, development, and protection; and free a large quantity of conserved water that could be used to meet current and future human and/or environmental needs in Mexico.

This and other similar proposals have identified a series of potential activities that could be associated with a binational effort to improve agricultural efficiencies.

- *Surface water and groundwater modeling in the border region:* At present, surface water and groundwater behavior and characteristics are poorly characterized in the border region. The development of a surface water model has been under discussion at IBWC for several years, but little progress on this model has been made to date, and in any event, an integrated model would be far more useful in terms of allowing for more informed decisions about the withdrawal of groundwater, the conjunctive use of surface water and groundwater, the construction of drainage facilities, and the design of environmental restoration plans.
- *Evaluations of water management:* The development of comprehensive water planning for the Mexicali Valley has been complicated by the delegation of operations and maintenance responsibilities to local districts, self-reporting strategies for the collection of data about water use, and limited information on the location and use of groundwater wells. A cooperative program could include analysis of irrigation performance and accounting practices on both sides of the border, identification of goals for improvement, and identification of meas-

ures to rehabilitate facilities, improve low-productivity farmland, address drainage and salinity issues, and ensure effective water accounting.

- *Irrigation improvements:* Observations of water use in the Mexicali Valley, particularly the central and western portions of the valley, suggest that efforts to increase system efficiencies and engage in conservation practices have not yet been as successful (or as well-funded) as corresponding efforts in the United States. As such, it is likely that there remain significant opportunities for improving water delivery and use through system automation, operational changes to improve the timing and quantity of deliveries, conversion to high-capacity farm turnouts, canal lining, spill interception, land leveling, installation of canal turnouts for rapid delivery, improved cropping patterns, changed field irrigation practices, adaptation to low-water technologies, and improved maintenance procedures. Water conserved from these efforts could provide replacement supplies in the face of shortages, thus reducing the dependence of local farmers on groundwater supplies and providing environmental benefits.
- *Drainage improvements in the Mexicali Valley:* Drainage improvements may provide an even greater opportunity for system enhancement than strides in water allocation and irrigation practices. In portions of the Mexicali Valley, canal seepage and deep percolation is pumped for irrigation, resulting in inevitable increases in groundwater salinity that place valuable croplands at risk in the absence of adequate drainage. Increased crop yields from drainage improvements could help offset socioeconomic impacts from the curtailment of other uses due to AAC lining or other developments, while making increased amounts of water available that could be dedicated to environmental uses.
- *Retirement of marginal agricultural lands:* Several recent studies, including a recent Packard Report, "Immediate Options for Augmenting Water Flows to the Colorado River Delta in Mexico," have investigated options related to taking existing, marginal agricultural lands in Mexico out of production and using the water from those lands for environmental purposes.

The Sonoran Institute and ProNatura Noroeste-Sonora, together with other non-governmental organization (NGO) partners, are currently in the process of exploring just such an option, focusing on highly marginal lands in the southern portions of the Mexicali Valley where salt buildup and shallow groundwater create economic challenges for agricultural production. Although funding for these efforts has not yet been secured, mechanisms have been identified for holding water derived from these lands via water trusts, wheeling water to appropriate locations, and designating protected receiving areas in the riparian corridor to ensure water is used for environmental benefits.

- *Improving municipal and industrial water supplies:* Mexico is currently experiencing an increased risk of shortages to municipal and industrial water supplies in the Mexicali Valley and the major communities to the west of the valley. Water conservation efforts, including conservation planning, promotion of conservation information and education, and implementation of conservation technologies and water management techniques, could help increase the availability of water for municipal and industrial uses.

These proposals were recently highlighted in a letter from the Sonoran Institute, ProNatura Sonora, Environmental Defense, the Pacific Institute, and Defenders of Wildlife to Secretary of State Condoleezza Rice, Secretary of Interior Gale Norton, and BOR Commissioner John Keys urging the United States to initiate discussions with Mexico on a broad range of binational water issues. The letter also suggested that other proposals involving only modest changes to the current international framework could increase flexibility related to water deliveries, result in net savings of water, make water available for multiple uses, or otherwise increase the availability of water in the border region for human and environmental uses. These include:

- *Water banking in Lake Mead:* Establishment of a top-reservoir water storage mechanism in Lake Mead, with corresponding changes in 1944 Water Treaty delivery obligations, could allow the accumulation and storage of water conserved through efficiency improvements, conservation efforts, fallowing programs, and other mechanisms to provide for periodic flood flows, water for environmental uses, or increased flexibility for water deliveries to U.S. and Mexican users in connection with temporary use reduction schemes.

- *Groundwater or off-stream wetland storage:* Investigations in the border region could reveal additional opportunities for groundwater banking of surplus flows in the United States and/or Mexico, as well as opportunities for temporary off-stream storage in Colorado River Delta wetlands, with potential benefits for water supplies and environmental resources.
- *Dry year options and drought contingency plans:* Provisions for binational dry year leasing or option programs could assist both countries with drought planning and the implementation of shortage sharing mechanisms. The recently released "Conservation Before Shortage" proposal for lower Colorado River basin shortage criteria (available at: http://www.environmentaldefense.org/documents/4601_CBSShortageProposal2.pdf) suggests there could be significant benefits associated with a program that would implement voluntary consumptive use reductions in the lower basin and Mexico as a method of limiting the extent and duration of shortages.
- *Use of administrative losses or regulatory augmentation:* Despite current proposals to reduce administrative losses through the construction of regulatory storage, it seems unlikely that administrative losses as a result of cancelled or reduced orders, small storm flows, and other unforeseen events will be fully eliminated. There may also be diminishing economic returns associated with limiting these losses beyond the current levels of control; as such, it might make more sense to commit these types of losses by agreement to be spilled out of Morelos Dam to support environmental uses in the limitrophe (potentially generating credits toward California's Multiple Species Conservation Plan) or for delivery via waste pathways to restoration areas in the delta.

- *Alternative delivery options to Mexico:* Demands for improved water quality for urban use in Mexico could be met through the implementation of alternative delivery paths for Mexico's 1944 Water Treaty entitlement, such as the construction of a new turnout on the AAC to reach users in Mexicali or the wheeling of water for urban use in Tijuana via facilities in Los Angeles and San Diego. The delivery of improved-quality water could be offset by the dedication of proportionately larger quantities of low-quality water for other uses.
- *Development or purchase of wastewater:* Significant opportunities remain in Mexico to improve the quality of water released to the environment via the construction or improvement of wastewater treatment facilities. Some facilities currently under construction may be able to supply water to support environmental uses. For example, the new plant in Mexicali, called Las Arenitas, could make water available to support environmental uses in the Río Hardy.
- *Changes to salinity requirements:* Salinity issues remain a concern for both Mexico and the United States. Mexico is concerned about periodic salinity spikes that impact agriculture in the Mexicali Valley, while the United States is concerned with the increasing difficulty of meeting the salinity differential developed in Minute 242 as overall salinity levels fall in response to successful salinity control programs. These problems may be complimentary. A more relaxed standard could potentially be negotiated to ease the costs of U.S. compliance in exchange for reducing spikes in salinity, reducing absolute salinity (such as increased investment in upper Colorado River basin salinity control efforts), investment in efficiency improvements, or other concessions.
- *Local and community economic impacts:* Many potential changes in water management, water allocation, and water use have the potential to generate economic impacts on individual water users and local communities. Economic adjustment funds to provide payments to offset economic impacts, redevelopment grants, retraining programs, and other measures could potentially help offset or prevent these types of impacts. These issues could prove to be a fruitful area for binational

discussion because third party economic impacts resulting from water management have been a significant issue on both sides of the border.

- *Yuma desalting plant workgroup recommendations:* A recent proposal developed by the Yuma Desalting Plant Ciénega de Santa Clara workgroup, organized by the Central Arizona Project, suggested a combination of adaptive management and environmental monitoring, a shortage mitigation trust fund, a basin-wide fallowing program, excess groundwater use in the Yuma area, and retasking of the Yuma Desalting Plant to meet potable water supply needs to ensure the continued supply of water to the Ciénega de Santa Clara while preventing impacts to U.S. water users.

In the final analysis, it may not be possible to preserve environmental resources such as the Andrade Mesa wetlands in their current condition. However, these and other solutions that have been proposed with regard to binational water issues could readily make available modest supplies of water to preserve, protect, and restore environmental values elsewhere in the Colorado River Delta while meeting critical socioeconomic needs in the border region. The good news is that, despite the growing threats to the environmental values in the delta, the amount (and quality) of water that would be required to preserve and restore these priority areas is surprisingly modest. A few of these priority conservation sites and the restoration opportunities and needs associated with them are highlighted below.

Colorado River Riparian Corridor below Morelos Dam

The Colorado River in Mexico extends approximately 153 km downstream from Morelos Dam to the Gulf of California. The first 96 km downstream from the dam are confined between flood-control levees, forming what is known as the riparian corridor. Varying in width from 1 km in the northernmost portion to 17.7 km in its lower end, the riparian corridor is subject to overbank flooding during major flood events. During the last two decades, low annual flows combined with large flood events have re-established significant native riparian plant communities.

According to the report "Conservation Priorities in the Delta" (Zamora Arroyo, et al. 2005c), this area is approximately 20,628 hectares in size and includes all natural areas within the levees (excluding agricultural lands) extending from Morelos Dam downstream to the point where the last stands of cottonwood and willow are found. The area includes the largest dense stands of cottonwoods and willows in the lower Colorado River basin. The area provides critical habitat for a variety of riparian birds and other wildlife species. In addition to protecting these existing areas, conservation and restoration actions could enhance the overall ecological integrity of the corridor by targeting areas within it dominated by bare soil and salt cedar stands.

Preservation and/or restoration needs include year-round base flow, occasional flood flows below Morelos Dam to sustain riparian habitat and marsh wetlands in the riparian corridor, non-native elimination, and native re-vegetation. Possible preservation/restoration actions include:

- Acquiring a 50,000 AF/y base flow through lease or purchase in the Mexicali Valley
- Acquiring water needed for a 260,000 AF periodic flood flow (3,900 cubic feet per second [ft3/s] for one-and-a-half months or 260,000 AF once every four years) over time through conservation actions, leases, or purchases of water and implementation of a banking agreement for storage at Lake Mead
- As instream flows are acquired, identifying funding for restoration actions, including planting native vegetation and channel improvements to reestablish hydrologic connectivity to the riparian corridor
- Developing a hydrologic model for the riparian corridor, performing site characterization studies, and developing a restoration design and implementation plan

Río Hardy Riparian Corridor

The Río Hardy is a perennial tributary of the Colorado River that carries approximately 6,000 AF to 11,000 AF of agricultural drainage water each year, flowing approximately 40 km from a point south of the Cerro Prieto geothermal facilities to the junction of the Río Hardy with the Colorado River near the tourist camp at Campo Flores. Local residents and NGOs are also working to protect an 17.7 km portion of the Colorado River downstream from the junction of the Colorado River and Río Hardy. Because the agricultural wastewater has high salinity (approximately 3,000 parts per million [ppm]), few native trees grow along the Río Hardy's banks. However, it remains an important corridor for water birds and songbirds. The portion of the Río Hardy corridor located within the Colorado River floodplain inundates during flood events and supports wetlands, shallow ponds, and hundreds of hectares of salt bush and salt cedar. Five restoration sites have been identified in the Río Hardy corridor.

Preservation and/or restoration needs include instream flows to sustain riparian forests and wetlands as well as native re-vegetation programs. Possible actions include:

- Dedicating, leasing, or purchasing up to 20,000 AF (28 ft^3/s) of treated effluent from the Mexicali II water treatment plant to provide instream flows in the Río Hardy (a minimum of 13,000 AF is required)
- Restoring vegetation by implementing a re-vegetation program with native plants and completing work on a check dam to maintain water levels, increasing the wetland area, and facilitating flows downstream into the estuary and inflows from tides into the river system
- Acquiring water and engaging in landscape modification to create additional wetlands, augmenting existing habitat areas, and serving as a buffer in case of malfunction of the wastewater treatment plant

El Indio Wetlands

This wetland restoration site is located on the southeastern end of the Mexicali Valley adjacent to the eastern flood control levee and within the Biosphere Reserve. The approximately 40.5 hectares of wetlands are maintained by agricultural drainage, which has created and maintained a marsh with areas of shallow open water and emergent vegetation, thus providing habitat for two endangered species, the Yuma clapper rail and the desert pupfish, as well as for waterfowl. For these reasons and for its high restoration potential, "Conservation Priorities in the Delta" (Zamora Arroyo, et al. 2005c) considers El Indio a priority for restoration. Restoration activities could potentially expand the wetlands to encompass approximately 779.8 hectares of surrounding low-lying areas.

Preservation and/or restoration needs include securing inflows to conserve and augment wetlands and improving inflow quality to broaden restoration options. Possible actions include:

- Obtaining a guarantee of 10,000 AF of agricultural drainage water to the site (securing the existing water source)
- Leasing or purchasing irrigation water to improve water quality and augment flows so that restoration options include cottonwood and willow forest in addition to mesquite and marsh wetlands.

Conclusion

Due to their significant and unique ecological value, particularly for bird species, to the Colorado River Delta ecosystem, the Andrade Mesa wetlands would ideally be protected and maintained in its current condition. Given the evidence that a hydrological connection may exist, it seems likely that these wetlands would disappear if the AAC lining project moves forward. As a result, additional analysis and mitigation for the project will likely be required under U.S. environmental laws.

The significant ecological value of the Andrade Mesa wetlands is due in part to its isolation provided by sand dunes, which offers greater protection against alteration and human impacts and enhances the ecological functions between wetland and desert shrub habitat. In addition, these wetlands are maintained by groundwater, which causes saturated soils that support particular vegetation species. Although there are areas in the Colorado River Delta where mitigation actions could take place, these characteristics make it difficult and costly to recreate the ecological values that will be lost if the Andrade Mesa wetlands disappear.

However, increased water demands in the Colorado River basin will inevitably increase the pressure on water managers to increase system efficiencies by engaging in projects like the AAC lining. To avoid a future that will be characterized by ongoing and protracted binational conflicts over water, it is essential that the United States and Mexico invest in an open process to resolve water allocation issues that will ensure the river will meet the human and environmental needs of both countries in the future, considering the ecoregional and binational context. The process already underway to develop shortage guidelines for the Colorado River could offer an excellent opportunity to begin a broader binational discussion of these issues. The need to reach a mutually agreeable interpretation of 1944 Water Treaty terms under which Mexico would share in a shortage with users in the United States creates an opportunity for the two countries to explore a broader range of options for the management of the Colorado River and its socioeconomic and environmental impacts.

In the context of growing pressure on water resources, binational issues related to the Colorado River cannot effectively be addressed in a piecemeal fashion. A more collaborative approach to water management could identify solutions that avoid zero-sum resolutions; offset the potential impacts of future shortage conditions on human communities and ecosystems in the United States and Mexico; help resolve ongoing disputes related to the AAC lining project, the proposed operation of the Yuma Desalting Plant, and other critical issues; and form the basis for improved binational cooperation and collaboration in the future.

ENDNOTES

[1] 40 CFR Part 1502.9(c).
[2] 40 C.F.R. § 1508.27. See also Warm Springs Dam Task Force v. Gribble, 621 F.2d 1017, 1024 (9th Cir. 1980).
[3] *Marsh v. Oregon Natural Res. Council*, 490 U.S. 360, 374 (1989).
[4] Bureau of Reclamation, *Revised and Updated Draft Environmental Impact Statement/Final Environmental Impact Report for Coachella Canal Lining Project* (2001).
[5] *See* 40 C.F.R. § 1508.27(b)(9). See also *Portland Audubon Society v. Babbitt*, 998 F.2d 705 (9th Cir. 1993) (court held that new information demonstrating a potential impact on an endangered species requires preparation of a supplemental EIS).
[6] *EEOC v. Arabian American Oil Co.*, 499 U.S. 244, 248 (1991); Foley Bros. v. Filardo, 336 U.S. 281, 284-85 (1949).
[7] Council on Environmental Quality, *Memorandum to Heads of Agencies on the Application of NEPA to Proposed Federal Actions of the United States with Transboundary Effects* (July 1, 1997).
[8] *See Natural Resources Defense Council v. U.S. Department of the Navy*, No. CV-01-07781 CAS(RZx), 2002 WL 32095131 (C.D. Cal. Sept. 17, 2002); Center for Biological Diversity v. National Science Foundation, 55 ERC 1873 (N.D. Cal. 2002); Province of Manitoba v. Norton, No. 02-cv-02057 (RMC) (D.D.C. Nov. 14, 2003).
[9] *C.f. Swinomish Tribal Community v. Federal Energy Regulatory Commission*, 627 F.2d 499 (D.C. Cir. 1980) (allowing Canadian intervenors to challenge the adequacy of an EIS regarding a dam project); *Wilderness Society v. Morton*, 463 F.2d 1261 (D.C. Cir. 1972) (allowing Canadian intervenors to challenge the adequacy of an EIS prepared for the trans-Alaskan pipeline); *Environmental Defense Fund v. Massey*, 986 F.2d 528 (D.C. Cir. 1993) (NEPA applies to National Science Foundation action in Antarctica).
[10] See *Trail Smelter Arbitration, U.S. v. Canada*, 3 UN Rep. Int'l Arbit. Awards 1911 (1941) (British Columbia had no right to operate smelter causing serious environmental harm in the United States); see also American Law Institute, Restatement of the Foreign Relations Law of the United States 3d, Section 601. This principle of customary law has been further recognized as a binding obligation of international law in Principle 21 of the Stockholm

Declaration on the Human Environment and Principle 2 of the 1992 Rio Declaration on Environment and Development, and includes a duty to give notice of anticipated impacts.

[11] 257 F. Supp. 2d 53, 66 (D.D.C. 2003).

[12] *Ibid.* at 66.

[13] *Ibid.* at 66–69.

[14] San Luis Rey Indian Water Rights Settlement Act, Pub. L. No. 100-675, § 203(a)(2), 102

References

Anderson, B. W., and R. D. Ohmart. 1984. *Vegetation Management Study for the Enhancement of Wildlife Along the Lower Colorado River.* A report by Arizona State University for the Bureau of Reclamation, Bureau of Reclamation Contract No. 7-07-30-V0009, July.

Comisión Nacional del Agua. 2005. *Efectos hidrlógicos que provoicará en territorio mexicano la recuperación del agua infiltrada en el Canal todo Americano, en California, Estados Unidos de América*. Subdirección general tecnica-Gerencia de Aguas Subterraneas.

Cortez Lara, A., and M. R. García Acevedo. 2000. "The Lining of the All-American Canal: The Forgotten Voices." *Natural Resources Journal* 40: 261–279.

Díaz Cabrera, P. 2001. "Simulación Numérica del Acuífero Superior del Valle de Mexicali, Baja California, México." Master's thesis, Centro de Investigación Científica y Educación Superior de Ensenada, Ensenada, Baja California.

Hinojosa Huerta O., H. Iturribarría Rojas, A. Calvo Fonseca, J. Butrón Méndez, and J. J. Butrón Rodríguez. 2004. *Caracterización de la Avifauna de los Humedales de la Mesa de Andrade, Baja California, México*. Final report of ProNatura, presented to the Instituto Nacional de Ecología. August.

U.S. Department of the Interior Bureau of Reclamation. 1994. *Final Environmental Impact Statement/Final Environmental Impact Report (FEIS/EIR), All-American Canal Lining Project.* March.

Zamora Arroyo F., P. Culp, M. Moreno, E. Santiago, O. Hinojosa Huerta, J. Butrón Méndez, M. Briggs, G. Anderson, and P. Titus. 2005a. *Valoración del Impacto Ambiental en México del Revestimiento del Canal Todo Americano y Alternativas de Mitigación. Fase I. Diagnóstico general del proyecto de revestimiento del Canal Todo Americano*. Sonoran Institute Special Report.

Zamora Arroyo F., P. Culp, M. Moreno, E. Santiago, O. Hinojosa Huerta, J. Butrón Méndez, M. Briggs, G. Anderson, and P. Titus. 2005b. *Valoración del Impacto Ambiental en México del Revestimiento del Canal Todo Americano y Alternativas de Mitigación. Fase II. Analisis de estrategias viables de mitigacion de los impactos en Mexico del revestimiento del Canal todo Americano*. Sonoran Institute Special Report.

Zamora Arroyo, F., J. Pitt, S. Cornelius, E. Glenn, O. Hinojosa Huerta, M. Moreno, J. García, P. Nagler, M. de la Garza, and I. Parra. 2005c. *Conservation Priorities in the Colorado River Delta, Mexico and the United States*. Report by the Sonoran Institute, Environmental Defense, University of Arizona, ProNatura Noroeste Dirección de Conservación Sonora, Centro de Investigación en Alimentación y Desarrollo, and World Wildlife Fund Gulf of California Program.

III

Fluctuations in Quality and Levels of Groundwater Near the Mexican-Proximate Portion of the All-American Canal

Jaime Herrera Barrientos, M. Norzagaray Campos, Gerardo García Saillé, Alfonso Andrés Cortez Lara, and Doris Jorquera Flores

INTRODUCTION

The All-American Canal (AAC) starts at the Imperial Dam 20 miles northeast of Yuma, Arizona, and is located on the ancient riverbed of the Colorado River. The canal moves 3,200 cubic hectometers (hm^3) of water per year (2.59 million acre-feet per year [MAF/y]). This water irrigates more than 200,000 hectares in California's Imperial Valley and supplies water to nine U.S. cities. The canal began operating in 1950, carrying water to the Imperial Valley. Its path brings it near the Mexican border adjacent to the Mexicali Valley in Baja California, where seepage occurs because the canal is not lined with concrete. Because the hydraulic gradient is oriented toward Mexico, the seepage from the AAC causes an elevation in the watertable that varies from 12.2 meters (m) to 24.4 m and feeds the Colorado River aquifer, and thus the Mexican portion of it in the

Mexicali Valley, with nearly 80.0 hm^3/y (65,000 AF/y) of the highest quality water in the valley's northeast. This chapter presents the fluctuations in the watertable and groundwater quality in the Mexican portion adjacent to the AAC, as well as the effect the canal's lining will have on Mexico's water supply.

Hydrogeology Associated with the All-American Canal

The AAC, in its 132-kilometer (km) (82-mile) path through the United States, has a 53-m (175-foot) downgrade. Its width varies from 46 m (150 feet) to 61 m (200 feet) and its depth from 2 m (6.5 feet) to 6 m (20 feet). Because it is not lined with concrete, water flowing through it seeps into the ground, providing water to the Yuma area (Hill 1996), the Imperial Valley in California, and the Mexicali Valley. Its path crosses through the Colorado River Delta and across the Mesa de Andrade, which is composed predominantly of sandy material with high permeability and is located adjacent to the Mexicali Valley. This terrain is one of the main sources of seepage toward Mexico. Figure 1 shows the AAC's location; note its proximity to the U.S.-Mexican border. The numbered points correspond to the location of wells that have watertable level series dating back to 1958.

The materials through which the canal passes are predominantly alluvial and have been deposited by the Colorado River throughout geologic time. Beneath the alluvial material are marine sediments of the Pliocene era and older, or volcanic and crystalline rocks. These materials constitute part of the Colorado aquifer, which extends from the United States to the Mexican portion in the Mexicali Valley. However, in reports from institutions such as the Comisión Nacional del Agua (CNA, in English, National Water Commission), the Mexican portion of the Colorado aquifer is identified simply as the Mexicali Valley aquifer. The thickness of the materials supplied by the river varies from 0 m at the foot of the mountains to more than 700 m (2,300 feet) southwest of Yuma (Gudel 1991).

Figure 1. Location of the All-American Canal and Pumping Wells

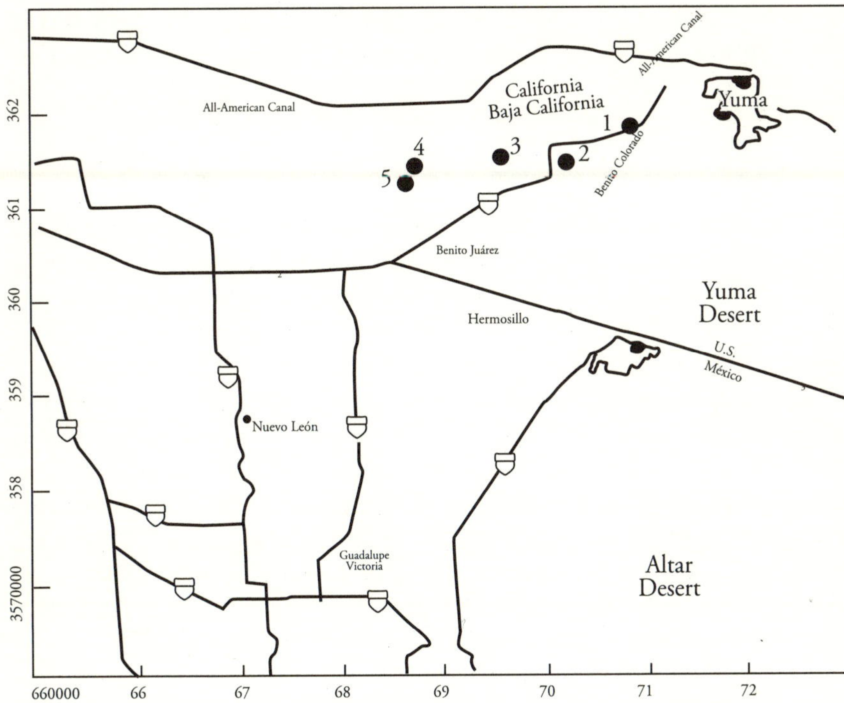

Notes: 1) 1-Ulloa, 2) 2-Cuervos, 3) 6-Mérida, 4) R-107, and 5) Borquez-8.

Hydraulic Characteristics of the Aquifer

The percolation, or seepage, capacity of the canal depends on the difference in the levels of water in the canal and the aquifer (otherwise known as the hydraulic load), the percolation rate, effective permeability, canal dimensions (length and width), and the hydraulic conductivity of the materials surrounding the canal's path. The mineralogy along the AAC's path consists of alluvial material deposited by the Colorado and Gila Rivers in the Yuma area. Perforations in wells in the valley (plateaus) of Yuma (Gudel 1991) have caused a pair of sedimentary bodies. The first (upper) has an average thickness of 60 m and consists of fine granular material above and coarse gravel in the deeper part, with sporadic insertions of thick sands. This gravel is referred to as the "thick gravel" area within which there are wells, both in Yuma and Mexicali, with excellent flow. The second body, below the first, reaches a thickness of some 600 m (2,000 feet) and consists of fine material in its upper part and grows coarser as the depth increases. Below these two alluvial bodies are sediments of marine origin as well as volcanic and crystalline rocks. This stratification pattern is common to both the U.S. and Mexican portions.

The first two bodies, with a thickness that varies between 60 m and 600 m, are of relevant geohydrological importance because of their permeability. It is here that most wells (more than 600 in the Yuma area and 725, both federal and private, in the Mexicali Valley) are located.

The mineralogical columns of wells such as 1-Ulloa, 7-Morelos, 1-Borquez, 9-Borquez, and 1-JoMaRodríguez, located at the northeast end of the Mexicali Valley adjacent to the AAC, show the same behavior described previously, with the gravel body growing deeper from east to west. The depth of the wells varies between 100 m (328 feet) and 180 m (591 feet), with 1-Borquez being the deepest. Noteworthy is the presence of layers, or strata, of clay interspersed in the gravel, especially in wells 7-Morelos and 1-Borquez.

Crossborder Flow

Various flow models take into account the Mexicali Valley aquifer, including CNA's bi-dimensional (horizontal flow) model from 1991, Hill's tri-dimensional model (1996), and Díaz's bi-dimensional model (2001). These models analyze the behavior of the rise in the Mexicali Valley's watertable, with the exception of Bradley's work, which extends it toward Yuma. CNA's model considers that, if seepage is reduced by 80% as a result of lining the AAC and the valley continues the consumption pattern of the last 10 years, water levels will decrease by nearly 6 m (20 feet) in the areas adjacent to the AAC. The effects would be noticeable 10 km (6 miles) away from the AAC in Mexican territory. In contrast, Díaz estimates that, should the AAC be lined, within 19 years water levels would drop below sea level in the areas of the valley with the greatest consumption. Hill (1996) analyzes the water supply sources responsible for the low depth of the water level in Yuma; likewise, it is estimated that if the AAC is lined in the Yuma portion, it would block the underground seepage of 138.15 hm^3/y (112,000 AF/y), leading to a reduction of the watertable adjacent to the AAC, within four years, of 2.8 m (9.2 feet).

Considering that the aquifer is homogenous and isotropic, the crossborder flow has been estimated taking into account the watertable series from the 40 years between 1958 and 1998. The 1958 series was chosen because it is the first one obtained after a total shutdown of well operations (pumping was suspended for 72 hours and the static level of the watertable then measured); the 1998 series was chosen because it reflects the behavior of the pressure gauge measurements from the last several years.

The crossborder flow (Domenico and Schwartz 1990) is a product of the hydraulic gradient (i), water conveyability (T), and length of the flow area (b). For the purposes of this calculation a section was drawn in the proximity of the U.S.-Mexican border and divided into regular, smaller-sized sections such that the length of all the sections equaled 34.16 km (21.22 miles), which corresponds to the portion measured from the point where the Colorado River crosses the east-west prolongation of the U.S.-Mexican border at Morelos Dam and the AAC's Drop 3. The values for T were compensated for in the calculations due to the variation in the watertable because the values for T correspond to 1972 (Díaz 2001). Being a free aquifer, they depend on the piezometric height, given that $T=Kb$, where the hydraulic conductivity K is considered constant and b is the saturated thickness, which changes over time in a free aquifer. The length of the subsections parallel to the border was selected considering the variation in T starting at $5x10^{-2}$ square meters per second (m^2/s). Tables 1 and 2 show the subsection data as median values of T, median hydraulic gradient, and T compensation factor for both the 1958 and the 1998 series. The first subsection (4.7 km) has been measured west from Morelos Dam. The T compensation factor was obtained by comparing the height of the watertable and T for 1972 (Díaz 2001) with the corresponding factors for 1958 and 1998. For example, if the piezometric height (*h*) and *T* for 1972 are 30 m and 11.67x106 m^2/y, respectively, then for *h*=24 m in 1998 the compensation factor is 0.8.

Table 1. Volume of Water Flow Considering the 1958 Series Data

b(km)	T(m^2/year)x10^6	Factor	i (m/km)	Q(hm^3/year)
4.70	11.67	1.00	1.7/6.31	14.77
2.10	10.10	1.00	2.0/6.31	6.72
2.10	8.51	1.01	1.0/2.63	6.86
2.63	6.94	0.91	2.0/5.80	5.72
2.10	5.36	0.96	2.0/5.80	3.72
2.10	4.73	0.96	2.0/5.80	3.28
2.63	2.21	0.96	2.0/3.68	3.03
15.80	1.57	1.09	4.0/5.26	20.56
34.16				64.66

Table 2. Volume of Water Flow Considering the 1998 Series Data

b(km)	T(m^2/year)x10^6	Factor	i (m/km)	Q(hm^3/year)
4.70	11.67	0.80	2.0/1.57	55.90
2.10	10.10	0.81	1.0/2.63	6.53
2.10	8.51	0.86	-1.0/5.26	-2.92
2.63	6.94	0.91	2.0/5.26	6.31
2.10	5.36	0.92	2.0/1.57	13.19
2.10	4.73	0.96	2.0/1.57	12.14
2.63	2.21	1.00	2.0/5.26	2.21
15.80	1.57	1.00	2.0/5.26	9.43
34.16				102.80

For 1958 (Figure 2), using the corresponding piezometric height series, the flow volume obtained is 64.66 hm^3/y (52,400 AF/y). For the 1998 series, shown in Figure 3, the flow volume is 102.8 hm^3/y (83,300 AF/y). The difference of 38.14 hm^3/y (30,900 AF/y) is due to the change in the hydraulic gradient because the 1998 value is greater than the 1958 value (as can be seen in Figures 2 and 3, as well as in the profiles in Figure 4). The change in the hydraulic gradient is induced by the reduction of the watertable level, a product of the relationship between the extraction of groundwater and seepage from the Colorado River, the AAC, crop fields, and irrigation canals, which constitute the net recharge of the aquifer.

Figure 2. Piezometrics for 1958

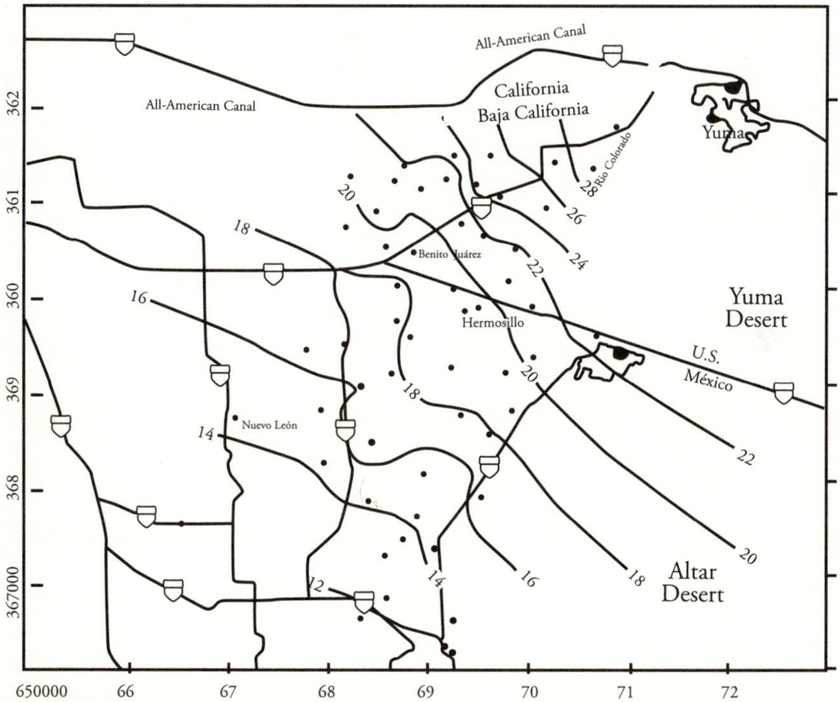

Note: Isocurves at 2 m. The points indicate the position of the pumping wells used for the preparation of the isocontours.

Figure 3. Piezometrics for 1998

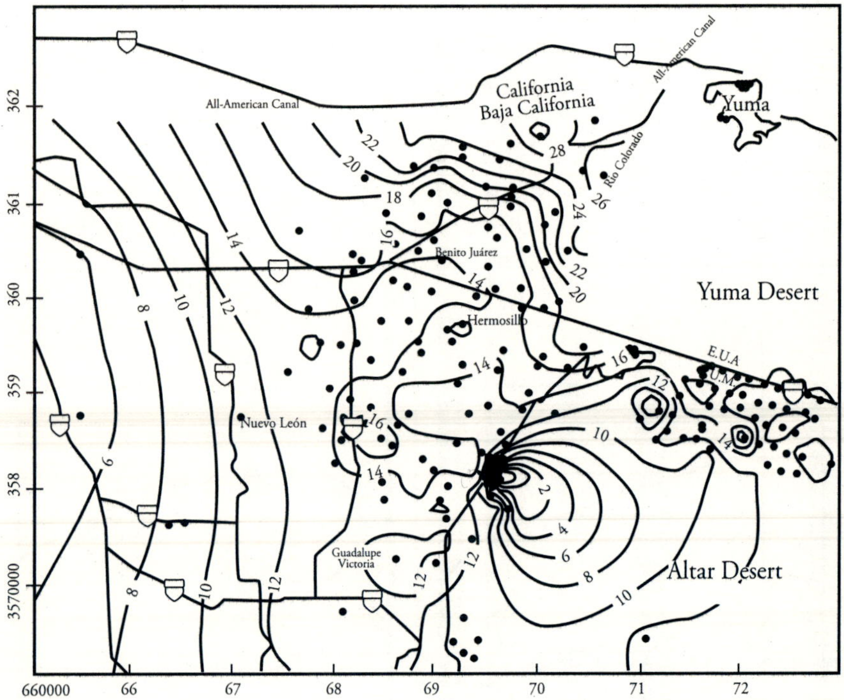

Note: Isocurves at 2 m. The points indicate the position of the pumping wells used for the preparation of the isocontours.

Figure 4. Watertable Level Profiles Taken from Morelos Dam Toward the Town of Benito Juárez

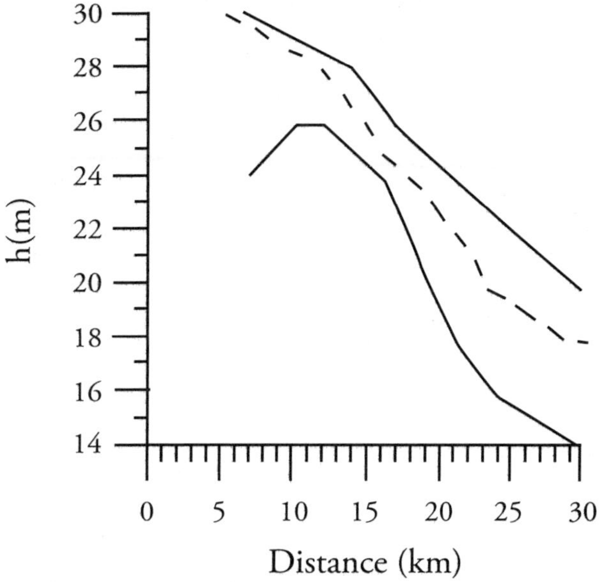

Note: The top solid line corresponds to the 1958 series, the dotted line corresponds to 1972, the bottom solid line corresponds to 1998.

The 102.8 hm^3/y (83,300 AF/y) of groundwater that enters Mexico is due to seepage from the AAC, Colorado River (normal flow and flooding), subterranean flow from Yuma, and agricultural activity. Therefore, considering that according to CNA (1991) the AAC supplies 80% of the water, then lining the AAC will reduce the underground flow into the Mexicali Valley by 82.4 hm^3/y (66,800 AF/y). This quantity varies from year to year due to the hydraulic gradient, which is a function of the relationship between the net aquifer recharge and the hydraulic load in the canal. This is why the areas next to the international border depend principally on the AAC. Figure 5 shows the height of the watertable for 1991; note that, as in Figures 2 and 3, near the Benito Juárez settlement there is a reduction in the hydraulic gradient reflected in Figure 4 as the area of convergence of the various series. This is interpreted as the limit of the area of direct influence of the AAC because the body of surface water behaves as a potential constant feeder of the aquifer.

Once past this area, the AAC contributes to the regional flow of the Mexicali Valley aquifer, in addition to the seepage from the Colorado River, recharge from irrigation, and the regional subterranean flow from the Yuma area.

The 82.4 hm³/y (66,000 AF/y) in seepage that would cease once the canal is lined represents 11% of the volume of extracted groundwater (because the annual median considered for the period from 1968 to 1994 is 750 hm³/y, or 608,000 AF/y) and 4.5% of surface water that Mexico receives from its agreement with the United States.

Figure 5. Piezometrics for 1991

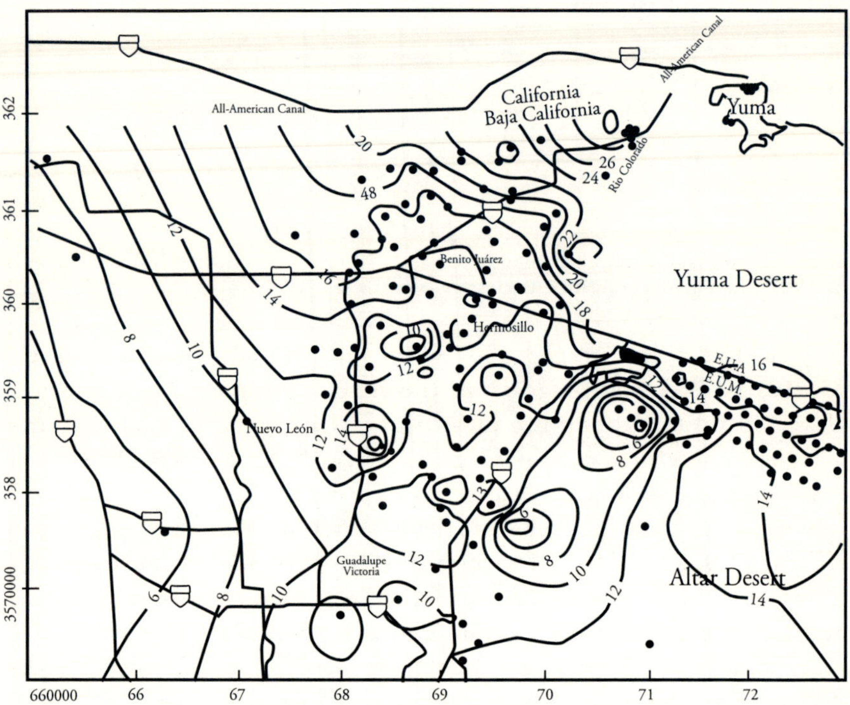

Note: Isocurves at 2 m. The points indicate the position of the pumping wells used for the preparation of the isocontours.

Evolution of Watertable Levels

The watertable level has decreased both spatially and temporally. In Figure 4, three profiles of pressure gauge measurements have X and Y zero-points that correspond to the area surrounding the Morelos Dam, with the X-axis passing through the Benito Juárez settlement. The curves do not start at the origin point because no data exist for that point. The thicker curve corresponds to the rise in the watertable in 1958, the dotted one to 1972, and the thin solid line to 1998. A reduction in the watertable levels can clearly be seen here. It increases over time such that the local reduction at the beginning of the 1998 series represents a total loss of 6 m when compared to the 1958 and 1972 series; it is also clear that the three curves converge at a distance of 17 km from the point of origin, after which they diverge. The convergence distance can be interpreted as the distance of influence of the AAC and is near the *ejido* Ciudad Morelos. It can also be seen in Figures 2 and 3 how the hydraulic gradients change. This is due, as has been indicated, to the rates of extraction from the aquifer and the magnitude of the different sources of seepage. According to the profiles in Figure 4, the spatial rate of watertable reduction is on the order of 60 centimeters per kilometer (cm/km), with the 1972 series (dotted line) reflecting the average conditions of the three series.

This downtrend is present in the previous profile as well as in temporal series in the following wells next to the AAC (see Figure 1 for their location): 1) 1-Ulloa, 2) 2-Cuervos, 3) 6-Mérida, 4) R-107, and 5) Borquez-8. The graphs of these five wells (see Figures 6a, 6b, 6c, 6d, and 6e) show a reduction in watertable levels over time, as seen in the flat sections obtained by a minimum squares adjustment method; the reduction gradients in each well are 10.0 cm/y, 7.0 cm/y, 7.0 cm/y, 12.0 cm/y, and 4.4 cm/y, respectively, the average being 7.3 cm/y, which indicates that the reduction over time and distance of the watertable is present in the vicinity of the AAC. Episodes of recharge associated with Colorado River surplus flows can also be observed. Their effect on the five wells shown begins in 1982, reaches its maximum in 1984, and ends in 1987. They then enter a process of abatement, which is also associated with Colorado River flow reduction. This spike disappears completely in 1990 because the flow associated with the surplus was 19,179 hm^3/y (15.548 MAF/y) in 1984 (Arellano 1992), which is the maximum for the period between 1972 and 1994. The flow in 1990 is 1,909.2 hm^3/y (1.547 MAF/y), or 10% of the total, likewise the volume of extracted groundwater in the valley in 1990 was the largest for the period from 1976 to 1994 (Díaz 2001). Colorado River surplus flows also appeared in 1993, with maximum flows of 700 cubic meters per second (m^3/s) (0.567 AF/s) in the March-to-May period, which resulted in the rise of watertable levels in the wells, as shown in Figure 5.

Groundwater Quality

This section addresses groundwater quality in terms of the concentration of total dissolved solids (TDS) in specific years. Regular sampling campaigns for water analysis in the Mexicali Valley began in 1984. These have measured Ph, electrical conductivity, basic cations (calcium, magnesium, and sodium), and anions (carbonates, bicarbonate, chlorides, and sulfates). Figure 7 shows the isocontours of equal values of concentrations of total dissolved solids measured in parts per million (ppm) or milligrams per liter (mg/l) for 1984. Note that, due to the extraordinary flow from the Colorado River that year (19,179 hm^3), the concentrations in the water next to the river are less than 1,000 mg/l, therefore the water can be classified as sweet or fresh water (Heath 1989), and in the rest of the valley concentrations are less than 3,000 mg/l and classified as slightly saline. In the Mexican portion next to the AAC, the concentrations vary between 900 mg/l and 1,300 mg/l, which classifies the water as between fresh and slightly saline.

Figures 6a, 6b, 6c, 6d, and 6e. Watertable Reduction Over Time in Pumping Wells

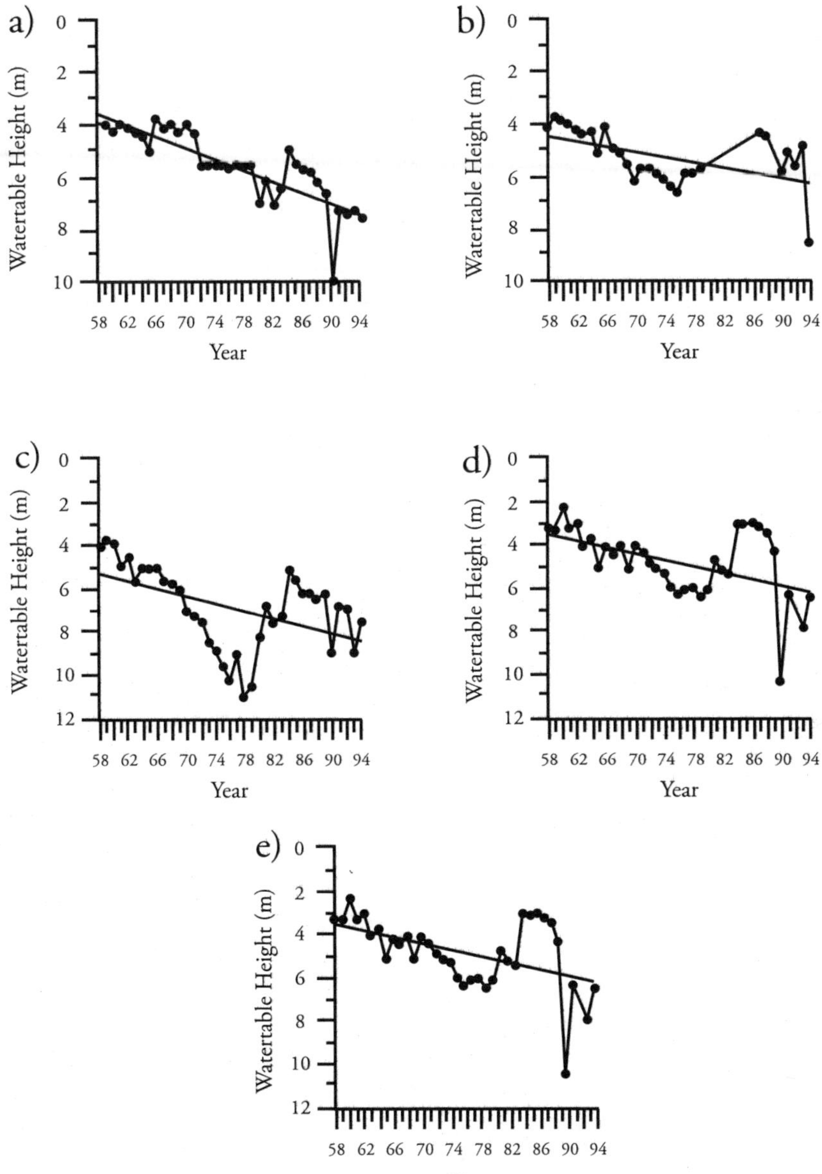

Note: a) 1-Ulloa, b) 2-Cuervo, c) 6-Mérida, d) R-107, and e) Borquez-8

Figure 7. Isocontours of Total Dissolved Solids (mg/l) in 1984

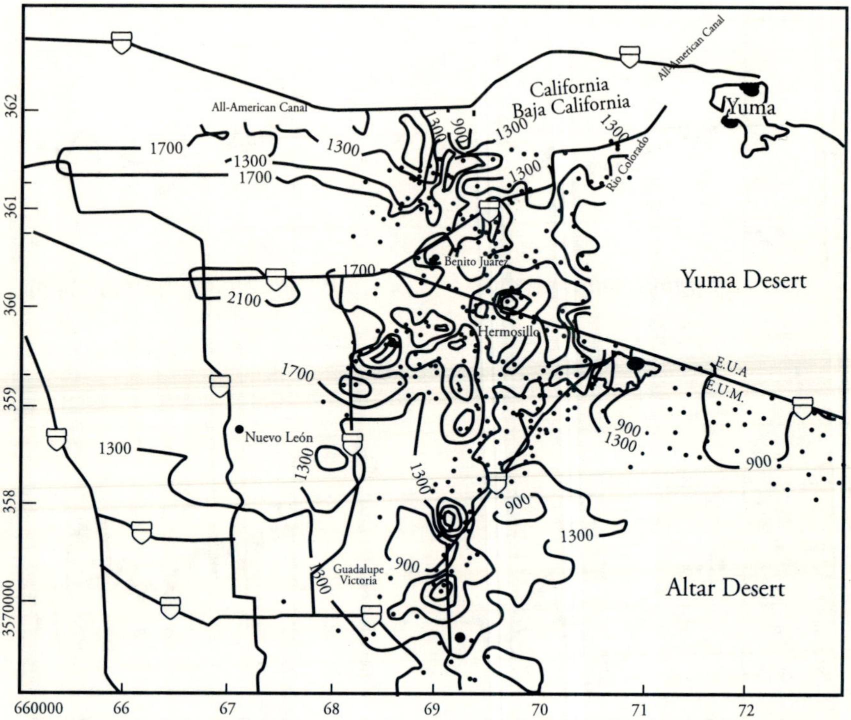

Note: The points indicate the position of the pumping wells used for the preparation of the map.

TDS concentrations in groundwater next to the AAC as it passes Yuma start at 600 mg/l (Gudel 1989). These concentrations can be observed northeast of Winterhaven, four miles from the Morelos Dam (Gudel 1989).

Figure 8 shows the isocontours of TDS concentrations for 1990. Note that in the proximity of the Colorado River the concentration has increased to 1,300 mg/l, which is greater than 1984's concentration, due to the fact that in 1990 the dissolution effects of the flow are less. Near the AAC, the concentrations vary between 900 mg/l and 1,300 mg/l, so there is no significant variation with respect to the series from 1984.

Figure 8. Isocontours of Total Dissolved Solids (mg/l) for 1990

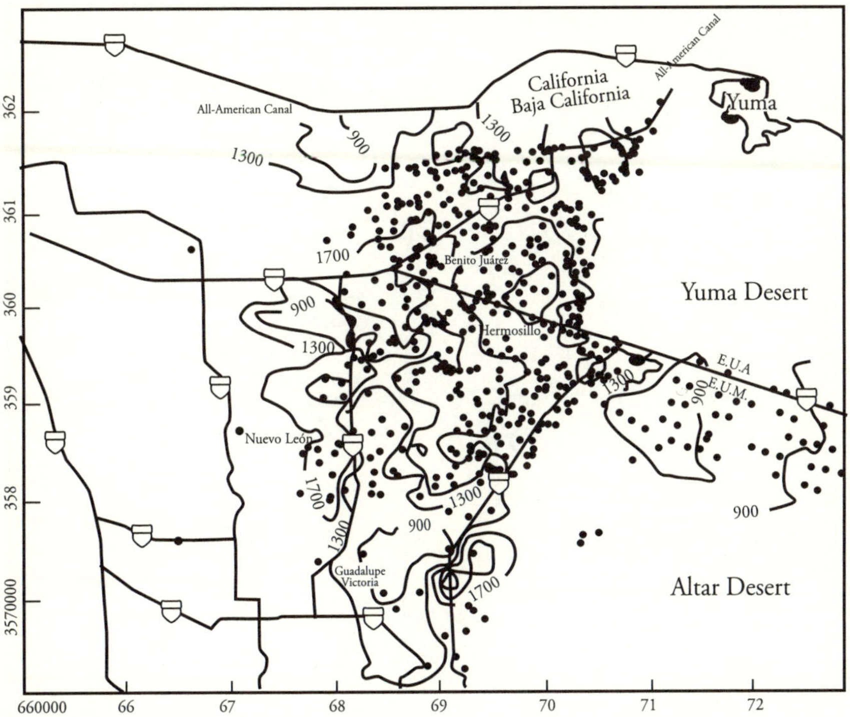

Note: The points indicate the position of the pumping wells used for the preparation of the map.

Figure 9 shows the isocontours for 1999. The permanence of the isocontours in the vicinity of the AAC can be seen, whereas in the rest of the valley the concentrations of TDS have increased noticeably compared to Figures 7 and 8, reaching concentrations of 2,900 mg/l west of Benito Juárez.

Figure 9. Isocontours of Total Dissolved Solids (mg/l) for 1999

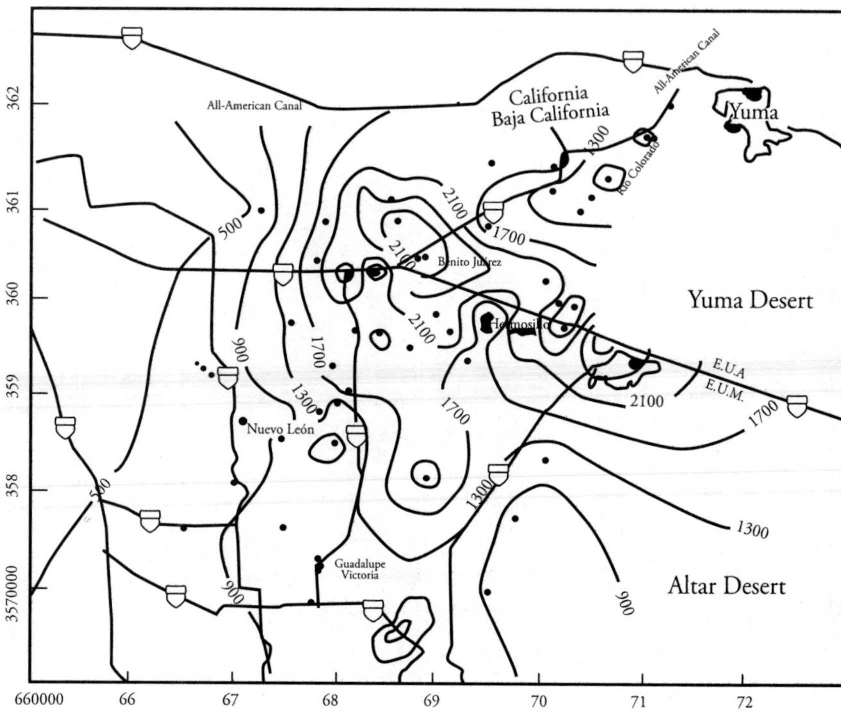

Note: The points indicate the position of the pumping wells used for the preparation of the map.

Figure 10 shows the isocontours for 2001. In the vicinity of the AAC the concentrations of TDS remain at 1,300 mg/l, maintaining the same pattern of the isocontours described previously.

Figure 10. Isocontours of Total Dissolved Solids (mg/l) for 2001

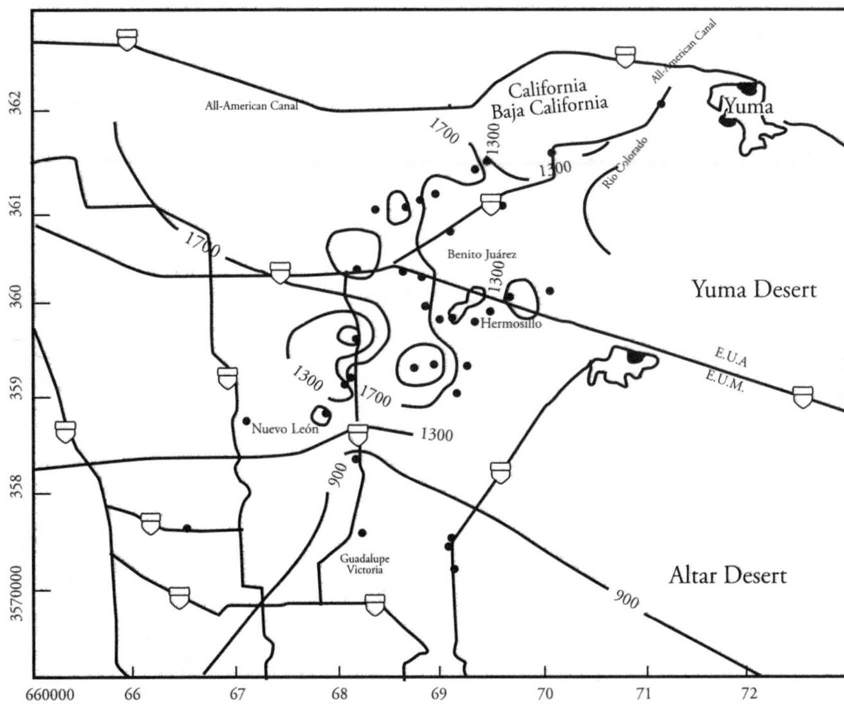

Note: The points indicate the position of the pumping wells used for the preparation of the map.

Conclusion

The AAC is a permanent source of good-quality groundwater that, once in the aquifer, has a concentration of TDS that fluctuates between 900 mg/l and 1,300 mg/l. This quality is maintained even taking into consideration periods of extraordinary supply and minimal flow from the Colorado River.

The annual volume of groundwater that enters Mexico through a 34.16 km (21.22 mile) section in the proximity of the AAC is calculated to be nearly 80 hm^3/y (64,857 AF/y). A reduction of watertable levels is observed, over both time and distance, northeast of the Mexicali Valley in the proximity of the AAC on the order of 7.3 cm/y (2.87 inches/y) and 60 cm/km (38 inches/mile), respectively.

References

Bear, J. 1972. *Dynamics of Fluids in Porous Media.* New York: Dover Publications, Inc.

Comisión Nacional del Agua. 1991. "Efectos del revestimiento del 'Canal Todo Americano' sobre territorio mexicano." http://www.cna.gob.mx/.

Díaz, C. P. 2001. "Simulación numérica del acuífero superior del Valle de Mexicali, Baja California, México." Master's Thesis, Centro de Investigación Científica y Estudios Superiores de Ensenada, Ensenada, Baja California.

Domenico, P. A., and F. W. Schwartz. 1990. *Physical and Chemical Hydrogeology.* Hoboken, N.J.: John Wiley & Sons.

Gudel, D. F. 1991. "Ground-Water Status Report. Yuma Area-Arizona, California." Yuma, Ariz.: U.S. Department of the Interior Bureau of Reclamation.

Heath, R. C. 1989. "Basic Ground-Water Hydrology." U.S. Geological Survey Water-Supply Paper 2220. http://permanent.access.gpo.gov/waterusgsgov/water.usgs.gov/pubs/wsp/wsp2220/index.htm.

Hill, B. M. 1996. "Use of a numerical model for management of shallow ground-water levels in the Yuma, Arizona area." *Ground Water* 34(3): 397–404.

Paredes, A. E. 1992. *Las aguas subterráneas en el Distrito de Riego 014, Río Colorado. Su Estado Actual.* Gerencia Estatal de Baja California, Internal CNA Report, Unpublished.

U.S. Department of the Interior, Bureau of Reclamation, Lower Colorado Region. 1994. "All-American Canal Lining Project, Imperial County, California: Final environmental impact statement/final environmental impact report." Boulder City, Nev.: Bureau of Reclamation.

Villanueva M., M. and A. Iglesias López. 1984. *Pozos y acuíferos. Técnicas de evaluación mediante ensayos de bombeo.* http://aguas.igme.es/igme/publica/libro35/lib35.htm.

IV

Lining the All-American Canal: Its Impact on Aquifer Water Quality and Crop Yield in Mexicali Valley

Gerardo García Saillé, Ángel López López, and

J. A. Navarro Urbina

Introduction

One of the main problems facing irrigated, arid areas is the availability of water for crops, and high salinity in the supply that is available aggravates this problem. The Mexicali Valley has two water sources to serve domestic, urban, industrial, and agricultural users in Baja California and part of Sonora, the greatest volume of which comes from the United States, whose source is the runoff into the Colorado River watershed.

The 1944 Water Treaty allocates 1.5 million acre-feet (AF) (1.85 billion cubic meters [m^3]) of water per year to Mexico. In addition, it takes 567,500 AF per year [AF/y] (700 million m^3) from an aquifer for irrigation and 159,711 AF/y (197 million m^3) for urban and industrial uses in the border cities of San Luis Río Colorado, Mexicali, Tecate, Tijuana, and Ensenada, as well as the rural population of the Mexicali Valley. The salinity from these sources has increased over time. Surface waters delivered to Mexico have steadily deteriorated, as suggested in studies by Comisión Nacional del Agua (CNA, in English National Water Commission), Universidad Autónoma de Baja California, and El Colegio de la Frontera Norte (COLEF).

Annual recharge of groundwater totals 567,500 AF. Of this volume, 81,000 AF seeps from the unlined All-American Canal (AAC). Some 16,216 AF of this is captured by the surface drainage system and is currently used in the irrigation of approximately 1,000 hectares (ha) (2,471 acres) of farmland. The remaining 74,864 AF are a recharge source for the aquifer.

The recharge from the AAC, in addition to being 14% of the volume extracted from the aquifer for irrigation, helps dilute the soluble salts in the aquifer. Approximately 70% of the recharge comes from over-irrigation (additional water applied to soil to avoid the accumulation of dissolved salts from the irrigation water itself) of Mexicali Valley crops with high-salinity water. The actual average of soluble salts in the aquifer is 1.8 grams per liter (g/L), and this increases by 20.6 milligrams per liter per year (mg/L/y). In contrast, the water incorporated into the aquifer from the AAC has a soluble salt concentration of 0.85 g/L and much lower annual increases than those in the Mexicali Valley aquifer.

The anticipated lining of the AAC will likely have two effects:

- A 14% reduction of the total available water in the Mexicali Valley, and a commensurate lowering of the static level
- An increase in the concentration of dissolved salts in the aquifer

Under these conditions, production of crops intolerant of salt will decrease, salts will accumulate in the soil, and a commensurate loss of productivity will result. Thus, growers will be forced to use more expensive technology or greater volumes of water to maintain production levels, either of which would result in a reduction of income per surface unit. Therefore, it is important to perform a quantitative and qualitative evaluation of the impact of lining the AAC on the water quality of the Mexicali Valley aquifer, the impact of this deterioration on crop production, and the loss of soil productivity due to its progressive salinization.

Background

In irrigated, arid areas, two of the main problems facing agricultural production are the availability and quality of water. The Mexicali Valley, located in the northwest corner of the country, is no exception. Agriculture has developed here only because of water from the Colorado River, although this has been controlled by a system of dams in the United States. The water treaty negotiated between the United States and Mexico in 1944 allocated to Mexico an annual volume of 1.5 million AF (1.85 billion m^3) from the Colorado River.

Various geohydrological studies have been carried out in this aquifer and suggest a median annual recharge of 567,500 AF/y, of which 405,403 AF are a result of over-irrigation during the crop-planting and harvesting process. Some 162,161 AF come from the periphery of the Mexicali Valley, 81,080 AF of this originates in the AAC at the northeast end of the valley.

The United States government intends to line the AAC with concrete to increase its conveyance efficiency and recover the volume lost to seepage in an approximately 53 kilometer (km) stretch starting at Pilot Knob. This chapter presents a general overview of the negative effects of this project on agricultural production.

The salt accumulation in the soil and the ensuing effect on crop yield are influenced not only by the volume and quality of irrigation water, but by weather, ground formation processes, stratification, and management, among other factors. These factors working alone or together could accelerate the loss of productive agricultural capacity. The following section describes the characteristics of the area where the most severe affects will occur as a result of reduced aquifer volume stemming from lining the AAC.

Area Characteristics

Geographic Location of the Area

The area expected to be most affected is the northeast section of the Mexicali Valley in the so-called first irrigation unit. It is physically occupied by irrigation modules 5 and 6 and bordered in the north by Mesa de Andrade, in the south by Federal Highway 2, in the east by the Colorado River, and in the west the agricultural area of irrigation module 13, which is incorporated into irrigation module 16. The largest towns located within this area are Vicente Guerrero, Los Algodones, and Ciudad Morelos "Cuervos." The Mesa Drain is located between the Mesa de Andrade and the agricultural area. The drain directly captures a significant proportion of the volume that seeps from the AAC, estimated at nearly 20,270 AF/y (Figure 1).

Figure 1. Location of the Possibly Affected Area in Irrigation District 014, Colorado River

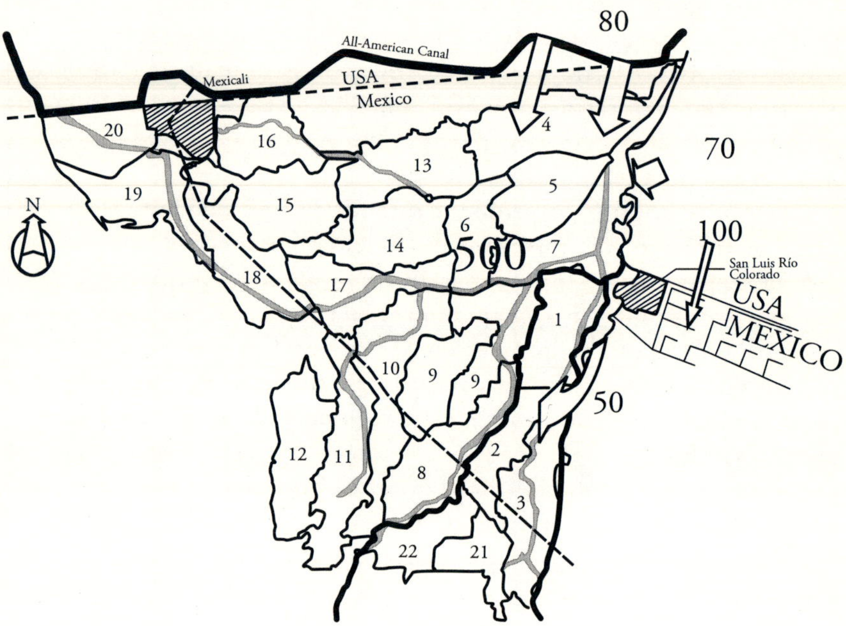

Topography

This area is nearly flat, with slight undulations and slopes of 2 centimeters (cm) to 3 cm per thousand centimeters with a generally northeast-southwest orientation. Maximum altitudes are approximately 40 m above sea level in the vicinity of Ejido Culiacán and minimum altitudes are 22 m above sea level in the western part near irrigation module 13. In the north-south orientation, altitudes measure between 32 m and 22 m above sea level.

Climate

The area's climate, according to the criteria established by García (1980), is classified as very dry and hot, with a median annual temperature of 22.3°C and extreme minimum temperatures of -7.7°C. Median annual precipitation measures 57 millimeters (mm) and potential evaporation is 247 cm. Under these conditions, salt accumulation processes in the soil are commonly present in the absence of irrigation, which is why permanent irrigation agriculture must be practiced.

Soil

The distribution of the different types of soil in the region, according to their mechanical composition, is presented in Table 1. Considering the characteristics of the series of soils, and grouping them according to similar characteristics in terms of ease of water handling and movement through them, it is clear that coarse soils occupy a surface area of 10,671.0 hectares (ha) (26,368.73 acres), sandy textures occupy a surface area of 7,468.6 ha (18,455.40 acres), and those with a clay texture 138.9 ha (343.27 acres). Coarse, sandy soil texture occupies 98.47% of the region's surface, which is an advantage for agricultural practices and the adequate development of crops. More than 95% of the soil is classified agriculturally as first- or second-class.

Table 1. Soil Distribution by Textures in the Affected Area

Series	Texture	Surface (hectares)	%
Gila light hase	Sandy loam	5,554.87	30.39
Gila heavy hase	Clay-lime loam	12,444.07	68.08
Imperial	Clay-lime loam	89.57	0.49
Holtville	Clayey loam	194.14	1.04
Total		18,262.65	100.00

Soil Salinity

Table 2 shows the soil classification distribution in terms of salinity and sodium parameters as outlined in a study of soil salinity carried out in 1995 by the office of Ingeniería de Riego y Drenaje (Irrigation and Drainage Engineering) of CNA. The table shows that soil salinity is not currently a serious problem for the development of agriculture in the area because just slightly more than 5% of the physical surface area with irrigation rights—some 1,007 ha (2,488 acres)—has moderately soluble salt accumulation problems. This contrasts with the general salinity conditions in the irrigation district soils that present soluble salt accumulation problems—nearly 50% of these soils range from slight to severe.

Water Availability

In the affected area, the main source of irrigation water is the aquifer. The non-profit Asociaciones Civiles, made up of the local irrigation modules, has the permission of the federal government to use 155,494 AF/y on a surface of 18,278.6 ha (45,167.41 acres). The irrigated area that uses gravity to bring water from the Morelos Dam is 277.9 ha (686.94 acres) and has an allocation of 2,792.42 AF. This provides a total surface area of 18,556.6 ha (45,854.35 acres) with an allocated volume of 158,286 AF/y, which means more than 98% of the water used in crop irrigation in the area comes from the aquifer. And in this area, seepage from the AAC is the most important contribution to the recharge of the aquifer.

Table 2. Surface Soil Classification in the Areas Affected by Salt Accumulations

Classification	Parameters		Surface	
	Electrical Conductivity	Percentage of Interchangeable Sodium	Hectares	%
1st Class	0-4	0-15	13,120.39	71.78
2nd Class	4-8	15-20	4,151.07	22.71
3rd Class	8-12	20-30	740.28	4.05
4th Class	12-20	30-40	266.86	1.46
Total			18,278.60	100.00

Current Irrigation Water Use

During the last few years the types of irrigated crops in the area have not varied, with the exception of the sesame seed crop, which was wiped out by a white fly infestation. In general, the crop surface tends to contract by always remaining above allocated rights. Table 3 shows the pattern of planted surfaces in the area from 1997 to date and Table 4 shows the gross water volumes (the total water extracted from the aquifer) per crop.

The information in Tables 3 and 4 establishes that the planted surface in the area has been greater than the surface with assigned irrigation rights and that the volumes used surpass the volumes assigned to the Asociaciones Civiles. This is due partly to the use of excess volumes for double crop planting, as well as to the acquisition of volumes from other *asociaciones* through irrigation permit transfers.

Evolution of Irrigation Water Quality

To evaluate the quality and degree of annual use of the aquifer, in the month of February wells are shut down for three days. Officials take readings on the third day for the static level of the aquifer and take samples for analysis of soluble salts and other parameters of agricultural importance, such as pH and electrical conductivity

Table 3. Planted Surface Pattern by Crop in the Area (acres)

Crop by Seasonal Type	1997-1998	1998-1999	1999-2000	2000-2001
Fall-Winter				
Safflower	54.73	2,007.96	398.11	0
Barley	80.67	3.24	31.62	0
Green onion	2085.47	2,359.52	1,206.48	880.54
Rye-grass	298.78	330.81	143.51	250.54
Wheat	62,965.77	52,128.28	85,900.07	67,060.18
Various winter	6,108.13	10,985.45	4,085.65	3,548.08
Subtotal	71,593.64	67,813.66	93,387.06	71,739.34
Spring-Summer				
Cotton	152,566.44	95,954.07	48,996.23	71,320.15
Sorghum early grain	31.62	1,696.29	780.81	114.31
Sorghum late grain	413.02	1,362.23	977.83	510.81
Sorghum early forage	1,363.78	3,527.49	5,115.38	2,571.87
Sorghum late forage	430.29	379.46	44.59	32.43
Early corn	168.97	466.29	2,139.73	676.21
Late corn	312.89	540.08	1,261.62	488.92
Various summer	129.73	690.16	658.37	318.65
Subtotal	154,605.94	104,616.08	59,974.54	76309.05
Perennials				
Alfalfa	42,298.23	36,535.99	26,145.26	25,696.08
Asparagus	8,607.12	8,961.84	9,827.78	9,917.78
Vine	1,678.61	1,118.91	1,065.40	745.13
Fruitages	639.65	414.48	377.84	287.84
Bermuda	0	0	0	0
Other perennials	0	0	0	0
Subtotal	53,223.61	47,031.22	37416.28	36,646.83

Table 3. continued

Crop by Seasonal Type	1997-1998	1998-1999	1999-2000	2000-2001
2nd Crops				
Sorghum grain	2,288.18	1,955.26	1,372.69	747.56
Sorghum forage	0	0	128.11	101.35
Corn	779.42	785.75	653.51	349.46
Various	184.22	41.35	9.73	0
Subtotal	3,251.82	2,782.36	2,164.04	1,198.37
Total	282,675.01	222,243.34	235,963	185,893.59

(EC). Some investigators, such as López L. (1991), using the information generated by CNA from these samplings as a baseline, have analyzed the behavior of soluble salt concentrations in the Mexicali Valley aquifer over time and distance. Among other things, these studies have determined that for each year of aquifer use, there is an average increase of soluble salts in the irrigation water of 21.8 mg/L. The foregoing is similar to the findings of a 1994 CNA study evaluating the years from 1961 to 1992; the study developed the following model:

$Y = 1{,}130 + 20.623\ (X)$

$R^2 = 0.993$

where,

Y = Concentration of soluble salts expected in the aquifer

X = Years, starting in 1961

R^2 = Determination coefficient

Table 4. Volumes Used by Crop in the Affected Area (acre/feet)

Crop	1997-1998	1998-1999	1999-2000	2000-2001
Fall-Winter				
Safflower	54.73	2,007.96	398.11	0
Barley	80.67	3.24	31.62	0
Green onion	2085.47	2,359.52	1,206.48	880.54
Rye-grass	298.78	330.81	143.51	250.54
Wheat	62,965.77	52,128.28	85,900.07	67,060.18
Various winter	6,108.13	10,985.45	4,085.65	3,548.08
Subtotal	71,593.64	67,813.66	93,387.06	71,739.34
Spring-Summer				
Cotton	152,566.44	95,954.07	48,996.23	71,320.15
Sorghum early grain	31.62	1,696.29	780.81	114.31
Sorghum late grain	413.02	1,362.23	977.83	510.81
Sorghum early forage	1,363.78	3,527.49	5,115.38	2,571.87
Sorghum late forage	430.29	379.46	44.59	32.43
Early corn	168.97	466.29	2,139.73	676.21
Late corn	312.89	540.08	1,261.62	488.92
Various summer	129.73	690.16	658.37	318.65
Subtotal	154,605.94	104,616.08	59,974.54	76309.05
Perennials				
Alfalfa	42,298.23	36,535.99	26,145.26	25,696.08
Asparagus	8,607.12	8,961.84	9,827.78	9,917.78
Vine	1,678.61	1,118.91	1,065.40	745.13
Fruitages	639.65	414.48	377.84	287.84
Bermuda	0	0	0	0
Other perennials	0	0	0	0
Subtotal	53,223.61	47,031.22	37416.28	36,646.83

Table 4. continued

Crop by Seasonal Type	1997-1998	1998-1999	1999-2000	2000-2001
2nd Crops				
Sorghum grain	2,288.18	1,955.26	1,372.69	747.56
Sorghum forage	0	0	128.11	101.35
Corn	779.42	785.75	653.51	349.46
Various	184.22	41.35	9.73	0
Subtotal	3,251.82	2,782.36	2,164.04	1,198.37
Total	282,675.01	222,243.34	235,963	185,893.59

Note: Data are from the agricultural statistics office of CNA's irrigation district 014.

Figure 2. Deterioration of Water Quality in Wells

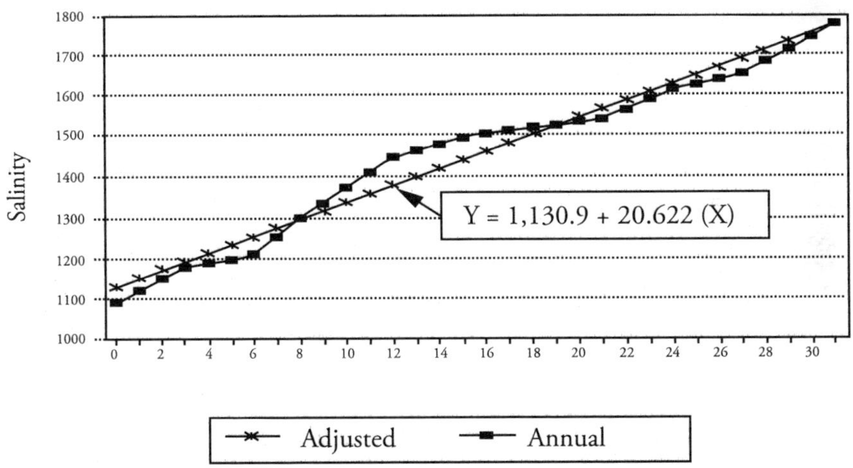

In contrast, Navarro (1998) analyzed the evolution of soluble salt concentrations only for water extracted from wells in the affected area. Navarro used similar methodology as López L. and CNA and arrived at the following model for aquifer water quality evolution in the area:

$Y = 1{,}179.84 + 21.929\ (X)$

$R^2 = 0.981$

where,

Y = Concentration of soluble salts expected in the aquifer

X = Years, starting in 1962

R^2 = Determination coefficient

These studies are summarized in Table 5. Note that the three studies carried out differ by less than 2 mg/L/y in terms of variations in the increase in soluble salt concentrations in the aquifer under current operating conditions.

Table 5. Relationship of Salt Increases Determined by Studies Conducted on the Aquifer

Author	Year	Model	R^2	Salt Increases (mg/y)
López	1991	No Model	-	21.8
CNA	1994	$Y = 1{,}130 + 20.623\ (X)$	0.993	20.623
Navarro	1998	$Y = 1{,}179.84 + 21.929\ (X)$	0.981	21.929

The largest proportion of water entering the aquifer—405,400 AF/y—comes from over-irrigation, that is, additional water that must be applied to the soil to avoid the accumulation of dissolved salts in the irrigation water itself. Both CNA, through its Office of Irrigation and Drainage Engineering, and COLEF (Cervantes and Bernal 1990) have studied the evolution of Colorado River water quality entering Mexico via the Morelos Dam. Both determined that soluble salt concentrations have increased at a rate of approximately 6 mg/L/y since the turn of the 20th century. In other words, the concentration of soluble salts in Colorado River water has increased from just more than 400 mg/L at the turn of the century to more than 900 mg/L at present. This evolution is presented in Figure 3.

The marked differences from the early 1960s to the late 1970s are due to the fact that Welton-Mohawk Irrigation and Drainage District sewage was incorporated into the Colorado River current. This sewage brought with it dissolved salt concentrations of up to 15,000 mg/L. It was mixed into the river's water before its arrival at Morelos Dam, and despite dilution in the river current, salt concentrations in water delivered to Mexico in that period for crop irrigation in the Mexicali Valley reached 2,500 mg/L.

The problems generated by the incorporation of sewage with high concentrations of soluble salts during this period were partially solved by both the United States and Mexico signing the International Boundary and Water Commission's (IBWC) Minute 242, but this did not address the effects this crop irrigation water had on the soil of the Mexicali Valley. The quality analysis of the water conveyed in the AAC and the volumes delivered to Mexico from 1990 to date are shown in Table 6.

Figure 3. Evolutionary Trend of Water Quality in the Colorado River and the All-American Canal from 1900 to 1990

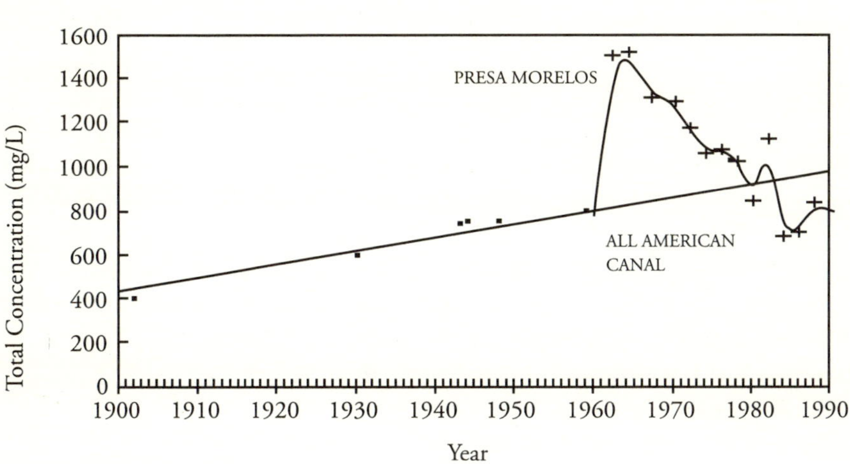

Source: Cervantes and Bernal 1990

Table 6. Annual Average Concentration of Soluble Salts in Water Conveyed in the All-American Canal and Delivered to Mexico at Morelos Dam (mg/L)

Year	Morelos Dam	All-American Canal	Difference
1990	910	775	135
1991	931	802	129
1992	965	836	129
1993	834	821	13
1994	965	872	93
1995	978	862	116
1996	982	834	148
1997	870	778	92
1998	785	714	71
1999	840	734	106
2000	850	726	124
2001	921	760	161
Averages	902.58	792.83	109.74

Thus, it can be concluded that the seepage from the AAC into the Mexicali Valley aquifer, and particularly into the area affected by the lining, contributes to the dilution of soluble salts in the aquifer.

Effects of AAC Lining on Water Quality

To generate a model that predicts future soluble salt concentrations in the Mexicali Valley aquifer, Navarro (1998) relies on the following assumptions:

- The Mexicali Valley aquifer has a total annual recharge of 567,500 AF/y, as determined in the Mexicali Valley geohydrologic study undertaken by the Secretaría de Recursos Hidráulicos (SRH, in English Secretariat of Hydraulic Resources) in 1972

- Of the total recharge volume, 64,860 AF/y is supplied by AAC seepage; this equals 31% of the total water used in the impact area, which has an allotted volume of 206,755 AF/y
- The remaining recharge in the area is composed of over-irrigation water, which totals 141,891 AF/y, or 69%, of the water used in this area
- The distribution of the 64,860 AF/y will be uniform in the affected area the first year after lining the canal, and the total aquifer recharge will be 503,000 AF/y

Given these considerations and performing the corresponding calculations, the anticipated concentrations of soluble salts in the aquifer were determined as shown in Table 7. This analysis establishes that, upon not receiving the 64,860 AF/y of seepage from the AAC, the concentration of soluble salts in the aquifer will increase from 1,879 mg/L to 2,004 mg/L, a difference of 125 mg/L, or more than five times what can be expected under current conditions. In subsequent years the concentration of soluble salts will increase by 23.5 mg/L.

Table 7. Evolution of the Concentration of Soluble Salts in the Affected Area After Canal Lining for 1-, 6-, 10-, and 20-Year Periods

Years After Lining	Model	Expected Concentration of Soluble Salts (mg/L)
1	C = 0.72 * 1873.29 + 0.28 * 2340.42	2,004
6	C = 0.72 * 1976.40 + 0.28 * 2489.65	2,120
10	C = 0.72 * 2058.89 + 0.28 * 2609.20	2,213
20	C = 0.72 * 2265.11 + 0.28 * 2908.07	2,445

Source: Adapted from Navarro 1998

Effect of the Changes in Water Quality on Crop Yield

The effect of salt concentrations on crops has been widely studied and described by many investigators, including Aceves 1979; Ayers and Westcott 1985; Fulton, Geattan, and Hanson 1993; U.S. Geological Survey 1985; Kaddah and Rhoades 1976; Kovda 1973; Mass and Hoffman 1976; Oster 1999; Rhoades 1990; Richards 1954; Shainberg and Oster 1978; and Wilcox and Durum 1974, among others. These teams investigated the relative tolerance of crops to soluble salt concentrations in irrigation water. At the same time, they have tried to relate the different levels of crop tolerance to crop yield under different salinity conditions and ionic ratios in the soil.

Investigations related to this same issue, but under local conditions, have been carried out by Cervantes 1983; Duarte, Rivera, and Russell 1987; Cisneros 1990; Cisneros 1993; Ruíz 1995; Palacios, Escamilla and Reyes 1978; García 1999; and Orozco 2001. These studies have demonstrated that a proportional response in yield reduction of different crop species exists when soluble salt concentrations in irrigation water occur.

Other authors have investigated the processes by which salts accumulate in the soil, including Jurinak and Suarez 1990, Kovda 1973 and 1980, Fetter 1988, Cervantes 1983, Svinarev and Bortsev 1962, Thornton 1981, Tanji 1990, Mendel and Shiftan 1981, and Voloubuyeb 1963. These projects establish methods that can be used to resolve salt accumulation problems in irrigated soil, in addition to the noted processes.

The impacts on water quality and availability in the Mexicali Valley as a consequence of the deterioration, both natural and induced, of aquifer water quality have been analyzed. Among these studies are Román 1991 and 1998, Cervantes and Bernal 1990, López 1991, CNA 1990 and 1995, and Navarro 1998. These analyze the impact of a degradation of surface and groundwater quality upon crop yield in the Mexicali Valley, whether by natural or induced causes.

Thus, to quantitatively evaluate the loss of this volume and the foreseeable increase in the concentration of soluble salts in the water, the decrease in crop yield in the area can be calculated by applying Mass and Hoffman's (1976) equations, which point out that salt tolerance is a relative value based on climactic and cultural conditions under which the crop is developed. Absolute tolerances that might predict the physiological responses inherent in the plants cannot be determined because there are many interactions between plants, soil, water, and environmental factors. These, in turn, influence the plants and their capacity to tolerate the presence of salts. The relative values that express the reduction in yield due to the effect of salts in the soil are described by the following equation:

$$Y = 100 - B(CEe - A)$$

where,

Y = Relative yield (%)

CEe = Average of soil salinity, expressed in EC of the soil saturation extract (deci Siemens per meter [dS/m])

A = Value of the conductivity where relative yield begins to decrease (dS/m)

B = Percentage of yield that diminishes per unit of increase in salt concentration

Under this model, one must first calculate the expected values of the EC of the soil saturation sample and then calculate the reduction in yield. These values are shown in Table 8.

Applying the relative yield equations for each crop, Navarro (1998) calculated the yield loss for the area's crops. These are presented in Table 9.

Table 8. Electrical Conductivity Values of Soil Saturation Extract Anticipated as an Effect of Lining the AAC

Year	Expected Concentration of Salts (mg/L)	CEa (dS/m)	CEe (dS/m)
1	2,004	2,592	3.9
6	2,120	2,742	4.1
10	2,213	2,862	4.3
20	2,445	3,162	4.7

Notes: CEa is EC of irrigation water; CEe is EC of the soil saturation extract
Source: Adapted from Navarro 1998

Table 9. Expected Relative Yield for Main Crops in the Area

Crop	% of Expected Production for Each Period (years)			
	1	6	10	20
Wheat	100.0	100.0	100.0	100.0
Barley	100.0	100.0	100.0	100.0
Rye-grass	100.0	100.0	100.0	100.0
Green onions	55.7	52.5	49.2	42.6
Various (veg.)	91.5	89.9	88.4	85.3
Alfalfa	86.4	85.0	83.2	80.7
Asparagus	86.4	85.0	83.5	80.7
Vine	76.8	74.8	72.9	69.0
Fruitages	66.2	63.0	59.8	53.3
Cotton	100.0	100.0	100.0	100.0
Sorghum G.T.	100.0	100.0	100.0	100.0
Various (summer veg.)	86.0	84.0	82.0	78.1

Table 9 suggests that the more tolerant crops—those with a better capacity for adaptation to higher concentrations of soluble salts in the soil—will not see their potential yield affected. However, more sensitive crops will suffer a considerable decrease in their total yield. One of the more notable cases is the crop yield reduction of green onions, which falls up to 58% over a 20-year period. For this crop in particular, a detailed analysis of the effects of AAC lining should be performed because the yield reduction will not only reduce the volume produced, and consequently the profit margin of the farmers, but it could have the collateral effect of increasing unemployment in the neighboring communities. This crop generates the largest need for labor in the area, providing employment for the day workers (jornaleros) who live in neighboring communities.

Additional Effects

To have access to these waters and maintain harmony in society, current legislation establishes "water right concessions." But, given the high demand for water in the region, nothing can be allocated in this manner. Likewise, the fact that lining the AAC will negatively affect the availability, quality, and pumping levels of groundwater, as well as the regional economy in one of the most productive areas of the irrigation district, it will result in reduced water supply for individual agricultural users.

An analysis of water rights revealed that currently, an irrigation volume of 115.2 cm is assigned to each hectare with recognized rights and is made up of 88.2 cm of Colorado River water and 27.0 cm of federal well water. To maintain balance in the aquifer recharge, lining the AAC would mean extracting 64,860 AF/y less, because federal well water availability would be reduced, given that in the affected area only 200 ha (494.2 acres) are irrigated with private wells. This means this volume reduction will be distributed equitably so that the total irrigation sheet will be 110.87 cm, composed of 88.19 cm from Colorado River water and 22.68 cm from federal well water.

The foregoing means an average reduction in the irrigation sheet per hectare of 4.29 cm for all the district's users. Approximately 6,943.85 ha (17,158.62 acres)—the surface area of one irrigation module—could be irrigated with that amount of water.

In terms of volume endowment and the water marketplace, which exists to balance the water rights with the actual consumption of the farmlands, users who have made technological improvements in their parcels will find their benefit annulled. Users farming high-consumption crops, upon having their allotment reduced, will need a higher volume of water.

Conclusion

Some conclusions can be reached based on this review of the available information. Under the established theoretical foundations with which the processes of conveyance and accumulation of soluble salts in the soil were studied, and given the effects of these accumulated salts on the growth and development of the crops (be it directly by the water itself, by the effect of individual ions in the internal metabolic processes of the plants, or by the deterioration of the soil's productive capacity), the processes of conveyance and accumulation of soluble salts in Colorado River water and the Mexicali Valley aquifer have suffered over time. A deterioration in quality has manifested in a steady concentration of soluble salts in the water. AAC seepage is useful not only for the volume of water supplied to the aquifer, but also because of the diluting effect on the concentration of soluble salts. Therefore, lining the AAC, or building a parallel lined canal, will have the immediate consequence of reducing the volume of water for the region as well as an immediate and medium-term increase of the concentration of soluble salts. This increase in soluble salts will result in a loss of 9% of the area's production and an increase of 13% in energy costs, which in turn constitute 25% of the operational and maintenance costs of the hydro-agricultural infrastructure of Irrigation District 014.

References

Aceves, N. E. 1979. *El ensalitramiento de los suelos bajo riego: Identificación, control, combate y adaptación*. Chapingo, México: Colegio de Postgraduados.

Ayers, R. S., and D. W. Westcott. 1985. "Water Quality for Agriculture." *Food and Agricultural Organization of the United Nations Irrigation and Drainage Paper* 29.

Cervantes, R. M. 1983. "La lixiviación de sales en suelos salinos y salinos-sódicos del Valle de Mexicali, durante un proceso de lavado." Master's Thesis. Colegio de Postgraduados, Chapingo, México.

Cervantes, R. M., and F. A. Bernal R. 1990. "Comportamiento de la salinidad en el agua del Río Colorado." In *Manejo ambientalmente adecuado del agua en la frontera México-Estados Unidos*, J. L. Trava Manzanilla, J. R. Calleros, and F. A. Bernal Rodríguez. Tijuana, B.C.: COLEF.

Cisneros, A. J. 1990. "Aprovechamiento del agua de drenaje agrícola para el riego de cultivos en el Valle de Mexicali." Master's Thesis. Facultad de Ciencias Agrícolas, Universidad Autónoma de Baja California, Baja California, México.

Cisneros, A. R. 1993. "Respuesta del cultivo de jitomate (*Lycopesicon esculentum Mill.)* a seis diferentes niveles de salinidad con relaciones iónicas cloruros-sulfatos variables." Master's Thesis. Universidad Autónoma de Baja California, Baja California, México.

Fetter, C. W., 1988. *Applied Hydrology*. Columbus, Ohio: Charles E. Merril Publishing Company.

Fulton, A., S. Grattan, and B. Hanson. 1993. *Agricultural Salinity and Drainage*. Davis, Calif.: University of California Irrigation Program, University of California.

García S., G. 1999. "Lindero Internacional Sur." Taller binacional sobre: Impacto de la salinidad del Río Colorado en los Valles de Mexicali e Imperial. Gobierno del Estado de Baja California, Universidad Autónoma de Baja California, University of California at Riverside.

García, E. 1980. "Apuntes de climatología." Report of the Instituto de Geografía, Universidad Nacional Autónoma de México.

Instituto Mexicano de Tecnología del Agua. 1994. "Anexo técnico, monitoreo de la calidad del agua de riego." Progreso, Jiutepec, Morelos. México.

Jurinak, J. J., and D. L. Suarez. 1990. "Agricultural Salinity Assessment and Management." *Manual and Reports on Engineering Practice* 71.

Kaddah, M. T., and J. D. Rhoades. 1976. "Salt and Water Balance in Imperial Valley, California." *Soil Science Society of America Journal* 40 (1).

Kovda, V. A. 1973. *Irrigation, Drainage and Salinity*. Rome: Food and Agriculture Organization of the United Nations.

Kovda, V. A. 1980. *Problem of Combating Salinization of Irrigated Soils*. Moscow: United Nations Environment Programme.

López, L., A. 1991. "Evolución de la Calidad del Agua del Acuífero del Valle de Mexicali, B. C." Master's Thesis. Instituto de Investigaciones en Agricultura y Ganadería, Universidad Autónoma de Baja California, Mexicali, Baja California, México.

Maas, E. V., and G. J. Hoffman. 1976. "Tolerancia de los cultivos a las sales." *Technical Bulletin* 9.

Navarro, U. J. A. 1998. "Impacto del revestimiento del All-American Canal en el Distrito de Riego 014, Río Colorado, Baja California y Sonora." Master's Thesis. Universidad Autónoma de Baja California, Mexicali, Baja California, México.

Mendel, S., and Z. L. Shiftan. 1981. *Groundwater Resources. Investigation and Development*. Burlington, Mass.: Academic Press.

Orozco, R. C. 2001. "Balance salino cualitativo del Distrito de Riego 014, Río Colorado." Master's Thesis. Universidad Autónoma de Baja California, Mexicali, Baja California, México.

Oster, J. D. 1999. "Proyecto Welton-Mohawk." Taller binacional sobre: Impacto de la salinidad del Río Colorado en los Valles de Mexicali e Imperial. Gobierno del Estado de Baja California, Universidad Autónoma de Baja California, University of California at Riverside. Mexicali, B. C. México.

Palacios, V. O., M. Escamilla, and A. Reyes. 1978. "El balance de sales del Distrito de Riego de Mexicali, B.C." *Natural Resources Journal 18*.

Rhoades, J. D. 1990. "Agricultural Salinity Assessment and Management." *Manual and Reports on Engineering Practice* 71.

Richards, L. A. 1954. "Diagnosis and improvement of saline and alkali soils." *U.S. Department of Agriculture Handbook* 60.

Trava Manzanilla, J. L., J. R. Calleros, and F. A. Bernal Rodríguez. 1991. *Manejo ambientalmente adecuado del agua en la frontera México-Estados Unidos*. Tijuana, B.C.: COLEF.

Román, C. J. 1998. Río Colorado: Agua en el Lindero Sur. Primer Congreso Internacional en Ciencias Agrícolas. Instituto de Ciencias Agrícolas, UABC, SAGAR, IMTA. Ejido Nuevo León, B.C. México.

Ruíz, C. J. S. 1995. "Aprovechamiento de agua residual tratada para Riego de Trigo *(Triticum vulgare L.)* en el Valle de Mexicali, B. C." Master's Thesis. Instituto de Ciencias Agrícolas, Universidad Autónoma de Baja California, Mexicali, B.C., México.

Secretaría de Recursos Hidráulicos. 1976. *Resumen del estudio geohidrológico del Valle de Mexicali, B.C. y Mesa Arenosa de San Luis, Son. México*. Mexicali: SRH.

Secretaría de Recursos Hidráulicos. 1972. *Resumen del estudio geohidrológico del Valle de Mexicali, B.C. y Mesa Arenosa de San Luis, Sonora, México*. Mexicali: SRH.

Secretaría de Fomento Agropecuario del Estado de Baja California. 1998. *Síntesis monográfica del Valle de Mexicali*. Mexicali: Secretaría de Fomento Agropecuario del Estado de Baja California.

Shainberg, I., and J. D. Oster. 1978. "Quality of Irrigation Water." *International Irrigation Information Center Publication No. 2*. Bet Dagan, Israel: Volcani Center.

Soleno, B. L. 1972. "Estudio general sobre el aprovechamiento de las aguas del Río Colorado, en el Valle de Mexicali, Baja California y San Luis, R.C., Sonora." Licenciatura Thesis. Escuela Superior de Agricultura Hermanos Escobar, Ciudad Juárez, Chihuahua, México.

Svinarev, V. I., and V. S. Bortsev. 1962. "Secondary Salinization of Kyzyl-Kum Desert Soils Under Irrigation with Artesian Waters and Methods of Eliminating It."

Tanji, K. K. 1990. “Agricultural Salinity Assessment and Management.” *Manual and Reports on Engineering Practice* 71.

Thornton, Iain. 1981. *Applied Environmental Geochemistry.* Burlington, Mass.: Academic Press.

U.S. Geological Survey. 1985. “Study and Interpretation of Chemical Characteristics of Natural Water-Supply.” Paper 2254. Washington, D.C.: U.S. Government Printing Office.

Voloubuyeb, V. R. 1963. *General Pattern of Changes in the Salt Content of Irrigated and Meliorated Soils.* Budapest: Institute for Soil Science and Agricultural Chemistry of the Hungarian Academy of Sciences.

Wilcox, L. V., and W. H. Durum. 1974. Quality of Irrigation Water. In *Irrigation of Agricultural Lands*, R. M. Hagan, H. R. Haise, and T. M. Edminster, eds. Madison, Wisc.: American Society of Agronomy.

V

The Colorado River and the All-American Canal: The Historical and Cultural Perspective of Water in the U.S. Southwest

Fernando A. Medina Robles

Introduction

When attempting to analyze a phenomenon, event, or social, political, or economic occurrence, the criteria applied have significant cultural content that corresponds to a particular cultural upbringing. This is not a problem, provided that the phenomenon defined and analyzed is within the context of one's own society and culture, whether local, regional, or national.

Everyone, in the end, is a product of their particular society and heirs of their historical upbringing and culture, which has formed a particular way of seeing things and interpreting events. However, in attempting to understand and analyze a phenomenon or event outside one's own culture or country, it is difficult to fully understand the causes or roots of the phenomenon, especially when trying to make sense of the behavior of the strangers participating in it.

How, then should a phenomenon be properly sized and analyzed? Clyde Kluckhohn suggests that comparative studies are a valuable resource. The study of societies different from an investigators' own:

> allows us to better see ourselves. Ordinarily, we are unaware of the special lenses through which we contemplate life. It could not be expected of investigators who did not go beyond the horizon of their own society to perceive the custom that is part of their own thinking. The man of science who studies human questions needs to know as much about the eye that sees, as about the object seen. Anthropology places before Man a great mirror and allows him to see himself in his infinite variety (Medina Robels 1970).

Thus, investigators must understand how those strangers see them and what categories of the strangers' own culture are used to form an image of who the investigator is. This approach allows understanding of attitudes and the foundation for decisions.

This chapter will examine the problem about to be generated for Mexico, and indeed the U.S.-Mexican border region, by a unilateral decision of the United States to line the All-American Canal (AAC) with concrete. This decision, with roots in the economic interests of the Imperial Irrigation District (IID), has generated protests on the Mexican side, and thus has become an international issue.

Much has been said and published on this issue on both sides of the border. One side justifies the lining project and the U.S. right to implement it, while the other justifies why it should not. The root of this controversy lies with the groundwater that is recharged from the canal's seepage and that is used by the farmers in Mexicali, Baja California, to irrigate their crops. Mexican producers would lose this water if the AAC is lined. Currently the issue is being discussed before the International Boundary and Water Commission (IBWC), a binational organization composed of representatives from Mexico and the United States who resolve the two countries' international water issues.

The intent of this chapter is not to establish who is right in this controversy, nor to examine opinions and reasoning on the controversy. The goal is to examine the historical processes that have formed water policy in the southwestern portion of the United States, the water culture generated in that process, and the people, political institutions, and private organizations that helped create the vision of water and its use in the extensive arid zone that is a significant part of the American Southwest. This vision constitutes the current American point of view Mexican negotiators must face. The adjustment of both visions of the problem will provide the elements to determine the present and possibly future decisions about the precious and scarce Colorado River water resource.

Unquestionably, water policy and culture in the U.S. Southwest—the result of an historical process of conflicts and adjustments between the seven states that use Colorado River water—has always affected, and will continue to affect, the development of the border region, which includes Baja California, whose only certain source of water is the Colorado River.

The Colorado River

Physical Aspects

The Colorado River is not considered one of the great rivers because its volume of water is relatively small. Rather, its importance lies in the fact that its extensive 244,000 square mile basin includes the territory of seven southwestern U.S. states: Wyoming, Colorado, Utah, New Mexico, Nevada, Arizona, and California. As well, the final 100 miles of its route enter Mexican territory and eventually discharge on an extensive delta in the Gulf of California.

Despite of its extensive tributary watershed, which includes the Salt and Gila Rivers, the Colorado River does not have a voluminous usable stream. It is only the sixth largest river in the United States because average annual precipitation in its basin is only 15 inches. Some 90% of its volume is lost through evaporation, and according to data collected from 1922 through the 1930s, it has a usable volume of 15.5 million acre-feet per year (MAF/y)—one-sixth the volume of water carried by the Columbia River. Considering these characteristics and the physical and climatic characteristics of the territories it traverses—and the characteristics of those that could potentially be rescued by its waters—it should come as no surprise that one of the river's historians, Norris Hundley Jr., states that as a consequence of the demands that have been and continue to be placed on its waters by its seven bordering states and Mexico, the Colorado River "has been one of the most litigated, regulated, politicized and discussed rivers in the world" (August 1999).

Water Policy and Its Effects on Water Culture in the American Southwest

The political, social, and economic leaders of its seven bordering states intervened in an historical process, representing the interests of the different groups that demanded their right to the specific use of water, according to their needs. As a consequence, a water policy and culture developed that sustained that situation. In this process, the figure of a Salt River Valley native in Arizona stands out: Senator Carl Hayden, who dedicated 67 years of his life in public service to efforts to establish policies for the development and use of Colorado River water.

Carl Hayden was not always a Senator. He began his incredibly productive political career at the age of 23, when he was elected in 1900 to the Salt River Valley City Council. In 1903 he was made a Captain in the National Guard. In 1904 he was elected treasurer of Maricopa County. He was elected county sheriff in 1907, a position he held for five years. In 1909 he was promoted to Major in the National Guard. In 1912 he won a seat as the first representative from the new state of Arizona. He was reelected to that office until 1927, when he was elected Senator, a seat he held until 1969 when, at the age of 92, he voluntarily retired. Hayden died in 1972 at age 94 (August 1999).

Among the various historic events that formed and guided the policies of development and use of Colorado River water, several stand out. On June 17, 1902, President Theodore Roosevelt signed the Progressive Newlands Reclamation Act. This act to reclaim lands by means of significant irrigation projects definitively established the policy that the federal government would undertake the irrigation necessary to rescue, on behalf of agriculture, vast arid regions in the U.S. Southwest to promote their population growth (August 1999).

On March 14, 1903, the Salt River Project was one of the first 26 reclamation projects authorized by the U.S. Department of the Interior under the conditions of the 1902 Theodore Roosevelt decree. Those 26 projects were part of the national irrigation program in its first decade (August 1999). The project was championed by Hayden and the leaders of the Salt River Valley Water Storage Committee, which later became the Salt River Valley Water Users Association. The users association prevails to this day as the generally accepted form of governance. The Salt River Project had great success in its development, a fact that allowed Hayden to establish a new policy with regard to the administration of federal reclamation projects that allowed the formation of irrigation districts, which in turn facilitated, many years later, the creation of IID.

Hayden always championed the concept of local control of federal reclamation works, even though they were of a public and national nature. In 1917, as a congressman, he persuaded the U.S. Secretary of the Interior, Franklin K. Lane, to transfer the "daily care, operation and administration" of the Salt River Project from the federal government to the Salt River Valley Water Users Association (August 1999). This was the first federal reclamation project to pass into the hands of its users, who would operate and administer them locally. Local control was so successful that it gave way to a new policy that became the blueprint for subsequent federal works throughout the southwestern U.S. (August 1999, Hundley 2002). It was further established, as a fundamental policy, that state residents would have the autonomy to determine the policies for preservation, conservation, and use of their natural resources.

The Great Controversy

When the exodus from east to west began in the United States and the pioneers of the Southwest established settlements, the settlers began the battle over the use of Colorado River water. The seven regions that made up the water basin later became federal territories and eventually part of the United States. Some of these regions first declared themselves republics, as did Texas and California. The exploitation of natural resources, especially mineral resources, created significant economic interests. Conflicts over water use between the states bordering the Colorado River arose to the same extent.

The farmers soon entered the controversy and insisted on defending their rights to water for agriculture, which in this vast semi-desert-like area had great potential for expansion and economic growth, if only the precious resource were transported to Nevada, Arizona, and California. Innumerable and lengthy lawsuits were filed between the states, some reaching as high as the U.S. Supreme Court. Among the best known are *Kansas v. Colorado* in 1907, *Wyoming v. Colorado* in 1922, and the longest-running and most expensive—11 years and $5 million—*Arizona v. California*, which began in 1952 and eventually ended when Arizona won in 1963. During this long controversy, which for all intents and purposes began in 1907 with the conflict between Kansas and Colorado, Hayden remained active, defending the rights of his home state to use Colorado River water. This ensured the distribution of water was equitable for the lower basin states, Nevada, Arizona, and California.

In November 1917, a California group established the organization League of the West. The organizers, with much political acumen, named Rufus B. Von Klein Smid, Chancellor of the University of Arizona, as the league's president. This greatly pleased the representatives of the Colorado River states, as well as the representatives from Oklahoma and Texas. During 1917 and 1918 the League of the West met several times in San Diego and Tucson, Arizona, and discussed issues such as economic development, tourism, commercial development, and transportation.

Part of the controversy arose between the upper basin states and Arizona and California. The other states were particularly concerned about the foreseeable development of Los Angeles and the Imperial Valley, which would demand considerable volumes of water. Hayden understood the concerns of the leaders of the upper basin states and their fears that, upon the rapid development of the lower basin, lawsuits would be filed based on prior and inequitable allocations of Colorado River water. Adding to these growing fears was the fact that the City of Los Angeles entered the competition for Colorado River water when it announced its need to ensure electricity and water for its rapidly expanding population. These fears were justified at a meeting of the League of the West held in August 1920, only four months after a harmonious meeting in Los Angeles. The representatives of the upper basin expressed serious doubts about California and its motives for the Colorado River. Oliver Shoup, Governor of Colorado, sounded the call to arms: "Now is not the time for the Western States that have the principal waters to lose any of their rights for any reason." His state engineer, A. J. McCune, added: "Our main fear is that Los Angeles and the people of the Imperial Valley commit the government to a policy that will interfere with our development" (August 1999).

The Colorado River Compact

These foregoing events convinced the political, social, and business interest group leaders in the seven Colorado River states that the most appropriate place for negotiations would be the U.S. Congress, and that the result could possibly achieve a more equitable distribution of Colorado River water among its users. The final result of all these conflicts was that, by 1920, Congress realized that the development of the Colorado River posed enormous and complex political, legal, engineering, economic, and diplomatic problems (August 1999).

To face these problems and provide an environment for negotiation to achieve the Colorado River Compact, president Warren G. Harding signed into law the Mondell Bill in August 1921 (August 1999). This act created the Colorado River Commission, which would be in charge of carrying out the necessary actions to create a draft Colorado River Compact acceptable to the seven states sharing the Colorado River Basin.

After holding public hearings in Washington, D.C., and meetings throughout the western United States, the commission prepared and signed the compact in November 1922 in New Mexico (August 1999). The stipulations and conditions of the compact divided the Colorado River Basin into upper and lower basins whose demarcation point would be Lee's Ferry, located in the desolate canyons of northern Arizona near the border with the state of Utah.

The commissioners assigned 7.5 MAF/y to the upper basin, composed of Wyoming, Colorado, Utah, and New Mexico. The Lower Basin, consisting of the states of Nevada, Arizona, and California, was assigned an equal amount from the main stream of the River plus an additional 1 MAF/y under the very ambiguously written and controversial Article III (*b*). According to the commissioners, water for agriculture and domestic use would have preference over power generation. The commissioners also almost completely ignored Mexico's right to use Colorado River water. In the event the United States should recognize that Mexico had rights to use of Colorado River water, Article III (*c*) stipulates that both basins would share the obligation in equal proportions (August 1999).

This blindness of the commissioners and their silence regarding Mexico's possible rights to Colorado River water echoed Hayden's position. In the summer of 1922, while the Colorado River Commission was in recess, Hayden informed the Committee on Irrigation of Arid Lands that:

> I do not want to see any treaty with Mexico until we have settled our own troubles in the United States. I shall oppose any kind of Mexican right in the Colorado River until it is definitely and fully determined there is a surplus of water in that stream for which there is no use in the United States (August 1999).

Carefully analyzing this quote in relation to the development currently achieved by California and Arizona reveals there has never been the possibility that a surplus would to be granted to Mexico as a permanent right. This clearly provides the perspective of the politicians of the Southwestern United States in terms of Mexico's water needs in the past and for the future. Hayden considered the Mexican issue a problem for the U.S. Department of State and the U.S. Congress, not for the Colorado River Commission.

In the entire historical process related to this controversy and the Colorado River Compact, Mexico had no active role in the defense of its rights, which were derived from the fact that the river discharges into a broad delta in the Gulf of California, within Mexican territory. As well, the river was navigable up to the port of Yuma, Arizona, and Mexico had to have control over vessel traffic that used Mexican territorial waters to access the river.

Therefore, the manner in which the Colorado River water reached the Imperial Valley, using the old riverbed of the Alamo River, whose approximately 50-mile extension is within Mexican territory, generates *de facto* the first rights of Mexico to Colorado River water. The possibility that Mexico would argue for rights to these waters was always a concern of Imperial Valley users, as they did not have complete control over the stream; it was also a strong incentive for the creation of their own irrigation district.

Mexicali Valley and Imperial Valley

The long historical process that led to the simultaneous beginning and joint development of the Valle de Mexicali (today the Municipality of Mexicali) and the Imperial Valley (today Imperial County) began long before the Colorado River Compact in 1922. It is rightly said that history is made by men. In the history of the Imperial and Mexicali Valleys, several men of vision and imagination intervened, men who early on saw the agricultural potential of the region, if only they could get Colorado River water to that extensive and fertile land.

In 1849, Dr. Oliver Wozencraft, on his way to San Francisco and the California gold mines, was the first to conceive of a plan for gravity irrigation in the Colorado Desert, later named the Imperial Valley, and thus develop an important agricultural and population center. Although Wozencraft dedicated his life to the project without finally achieving it, his frustrated attempt established the basis for future efforts.

Wozencraft began his stubborn task of bringing together the necessary capital for his reclamation project in 1853. He contracted the services of Ebenezer Hadley, then Los Angeles County Surveyor, who recommended locating a canal to divert Colorado River water at a point that, 40 years later, was designated as the adequate place to begin the project, "a point of rocks adjacent to Pilot Knob immediately above the International Border" (Dowd 1956). Hadley noted on that occasion the canal would need to pass through Mexican territory.

Wozencraft managed to obtain the support of the California State Legislature in the approval of a bill on April 15, 1859, wherein the state granted "all rights, title and interest, of the State of California, which the State may have at the time, or those it may acquire in the future, on and over the lands described herein." The land described in the bill included, in general, everything that is today the Imperial and Coachella Valleys. The concession was granted on the condition that within three years of the approval of a bill in the U.S. Congress granting these lands to California, Wozencraft would provide "a permanent supply of fresh and healthy water along the line of travel between the San Gorgona (*sic*) Pass and Fort Yuma, and Cariso (sic) Stream and Fort Yuma, in such fashion as ensuring there be a constant and permanent supply, in all seasons and convenient points necessary on the two routes mentioned above" (Dowd 1956). That same year, Congress considered a resolution that nearly 3 million acres of land in the Colorado Desert be transferred to California for reclamation and irrigation. Wozencraft remained in Washington many years to ensure some action by Congress on his plan; his efforts did not end until his death in 1887 (Dowd 1956).

In 1879, John C. Fremont, a great promoter and at the time interim governor of the territory of Arizona, proposed a project that greatly upset his constituents on the banks of the Colorado River. Fremont supported a plan to alter the climate of the Imperial Valley by flooding the Salton Sink (today the Salton Sea) with Colorado River water. Arizona residents vigorously opposed the measure.

In 1892, another 19th century pioneer arrived in the region, who finally achieved Wozencraft's dream. Engineer Charles Robinson Rockwood arrived in Yuma in September 1892, having been hired as chief engineer by the Arizona and Sonora Land and Irrigation Company of Denver. He was accompanied by C. N. Perry and a team of engineers. The purpose of the trip was to perform a survey to explore the possibilities of building an intake to channel Colorado River water to irrigate and develop 1.5 million acres the company owned in Sonora (Heffernan 1930).

While investigating the possibility of irrigating lands along the border, Rockwood observed, as had Wozencraft before him, that the Imperial Valley could become an extensive, useful garden year-round if water were brought to its fertile lands.

> In the meantime, however, while these surveys were in progress, I had taken a team and made a trip into that portion of the Colorado Desert, known today as the Imperial Valley. The result of my investigation at that time was such that it led me to believe that, without a doubt, one of the most worthwhile irrigation projects in the country would consist of bringing together the lands of the Colorado Desert and the waters of the Colorado River (Rockwood 1930).

Rockwood sent to the Denver company a preliminary report indicating that the Sonora project was not economically feasible because the area of farmable land did not guarantee recovery of the enormous investment required (Heffernan 1930). However, Rockwood, in that same report written in early 1893, recommended performing surveys to prove or disprove his belief in the merits of the project for the Colorado Desert. He was authorized to run preliminary parallel lines to determine levels, acreage of land that could be reclaimed and made available for agriculture, and the approximate cost of construction (Rockwood 1930).

The response from Denver was so enthusiastic that officials there immediately took steps to change the company's name from Arizona & Sonora Land Irrigation Company to Colorado River Irrigation Company, and they maintained that they had $2 million available to undertake the project if the survey results were favorable (Rockwood 1930). With these assurances, Rockwood began making his vision of the Imperial Valley a reality. The path was plagued with frustrations, owing to financial problems and legal obstacles within the United States as well as internationally, because any diversion of Colorado River water to the Imperial Valley had to inevitably cross through Mexican territory.

The first problem arose in 1893 when the directors of the Colorado River Irrigation Company suffered financial setbacks to such a degree that they were unable to continue with the project despite the results of the surveys, which were even more favorable than initially estimated (Rockwood 1930). For Rockwood's good fortune, during his stay in Yuma in 1892 he met Dr. William T. Heffernan, who was employed by the U.S. government as a surgeon there. They established a great friendship and mutual appreciation that lasted the rest their lives (Heffernan 1930). Very soon Heffernan made Rockwood's project his as well, to the point of providing on several occasions the necessary funds to continue it.

By the time the Denver company abandoned the project, Rockwood had paid company bills out of his own pocket and ultimately had to cover personnel salaries as well. Out of resources and with no hope of recouping the money he had paid out on behalf of the company, in addition to several months of his own salary he had not been paid, he sued the Denver company and obtained a lien against all their Yuma assets, including engineering equipment, reports, and logs of the surveys performed. When these assets were auctioned, Heffernan acquired the entire lot for $7,800. This became Heffernan's first investment in the Imperial Valley reclamation project (Heffernan 1930).

Because of the problems surrounding the Colorado River Irrigation Company, Rockwood and his associates, Heffernan among them, set up the California Development Company in New Jersey on April 25, 1896; Rockwood was named vice-president (Dowd 1956).

According to its bylaws, this company was authorized to build and operate irrigation systems; store and distribute water on its own behalf as well as that of third parties; buy and sell real estate both in the United States and in Mexico; buy and sell water rights; establish settlements; build roads, railroads, and other means of transportation; lay telegraph and telephone lines; build power plants and water treatment utilities; and establish cities. In 1899, Rockwood was named president of the company and remained in that office until 1900, when George Chaffey was hired as president (Dowd 1956).

As a result of negotiations with Chaffey, he took charge of the project and under the terms of a contract signed in April 1900, he was granted full control of the California Development Company and its Mexican subsidiary. In exchange, Chaffey would finance the head-works and construction of the canal that would bring Colorado River water to the Imperial Valley. The total cost would not exceed $150,000 and the canal would have a water flow delivery capacity of 400,000 AF/y. Chaffey would be reimbursed for the cost of construction in addition to $60,000 as compensation for his services for five years (Dowd 1956).

Finally, at 11 a.m., on the morning of May 14, 1901, the first Colorado River water was diverted to the Alamo River and, passing through the Mexicali Valley in Mexican territory, arrived in the Imperial Valley on June 21, 1901 (Gobierno del Estado de Baja California 1958). However, international legal problems, questions concerning the control of the water diversion route south of the border, problems related to the Mexican Revolution, and the presence of American speculators in Mexican territory—who frequently operated against the interests of the Imperial Valley—generated a series of crises for Rockwood, the Colorado Development Company, and the residents of the Imperial Valley itself (August 1999).

The irregular manner in which the Alamo River was used in the diversion of Colorado River water to the Imperial Valley caused several problems for the project's promoters. The land through which the Alamo River flowed was within Mexican territory and was the property of a Mexican citizen from whom the land had to be purchased. Later on, another problem arose when foreign companies were legally prohibited from owning land in Mexico, as were operating companies that provided public services such as delivery of water.

To resolve these jurisdictional problems, two companies were formed: the California Development Company in the Imperial Valley and the Sociedad de Irrigación y Terrenos de Baja California in the Mexicali Valley (Dowd 1956). Thus, the U.S. company signed the necessary contracts with the Mexican company to facilitate the conveyance of Colorado River water and its delivery to Imperial Valley users. On December 28, 1900, a contract was signed in which the California Development Company committed to deliver to the Sociedad de Irrigación y Terrenos de Baja California, at Hanlon Heading on the U.S.-Mexican border, sufficient water to allow the Mexican company to supply the lands of the Imperial Valley and those in Mexico. In addition, the California Development Company agreed to build, operate, and maintain the principal canal system on the Mexican side of the border.

The California Development Company continued to have operating problems, such as inefficiency in delivering water to Imperial Valley users. There were also serious financial problems and lawsuits against both the Mexican and U.S. companies, including those due to the lack of proper operating authorizations and concessions that had to be granted by the Mexican government for the operation of the Mexican company. In truth, this company was a public service utility and the Mexican government had not granted the necessary authorization. The concession was granted on May 17, 1904, by an act of the Mexican Congress and signed by the president on June 7, 1904. That same concession authorized the Mexican company to divert, within Mexican territory, a volume of 284 cubic meters per second of Colorado River water through its canal system, provided neither third party interests or navigation were affected. The company was authorized to send the waters transported through its canals to the Imperial Valley for irrigation, provided sufficient volume was set aside to service irrigable lands in Baja California through this system, without exceeding one half of the volume of water that should pass through the canals. Later on, the Mexican government would interpret this provision in the concession to mean one half of the volume flowing at any given time, after deducting the amounts necessary for regulation and seepage in Mexico (Dowd 1956).

The Imperial Irrigation District (IID)

What finally determined the financial ruin of the California Development Company was a series of Colorado River floods in 1904 and 1905. The Gila River overran its banks and expenses incurred to stop the flood that resulted were covered by loans from the Southern Pacific Railroad Company. The California Development Company was unable to pay back the loans, and this later led to a takeover by the railroad company. Once the flood was controlled, though, the California Development Company was subjected to innumerable lawsuits for damages and payment of past-due debts. The Sociedad de Irrigación y Terrenos de Baja California was facing settlement of a lawsuit totaling 900,000 pesos in gold. There was no doubt that both companies were bankrupt. To satisfy the payment of the lawsuit brought by the Southern Pacific Railroad against the Sociedad, a new Mexican company was established, the Compañía de Terrenos y Aguas de Baja California, which acquired all the assets of the Sociedad in a sale closed on January 28, 1911. The price was 325,000 pesos in gold, approximately 36% of the amount owed (Dowd 1956).

This sale left the old Mexican company without assets with which to respond to the remaining debt of 375,000 pesos in gold and new lawsuits for damages, which increased its debt to 2 million pesos. The sale also freed the new Mexican company of all contracts with the California Development Company and the water users' mutual companies that had organized in the Imperial Valley. This could have given the new Mexican company total control over the development of the Imperial Valley (Dowd 1956).

This legal maneuver isolated the problem in Mexico and, in a way, allowed the irrigation system in Mexicali to continue operating and support the Imperial Valley's system. This was only a reprieve, though, because the California Development Company continued to have financial problems not only with the Southern California company but with other creditors as well. This chain of problems generated discontent and uncertainty among the residents of the Imperial Valley and on July 14, 1911, they voted to create IID (Dowd 1956).

The creation of the irrigation district in no way resolved the serious problems of past-due debts of the California Development Company or of the two Mexican companies, the Asociación de Irrigación y Terrenos de Baja California and the Compañía de Terrenos y Aguas de Baja California. In the meantime, the Southern Pacific Company had acquired all rights to the debts, which finally made it possible to resolve in favor of IID the assets of the companies in conflict. IID, through a bond issue, obtained the necessary funds to buy the Southern Pacific Company's interest in all the assets, which consisted of the canal systems, water rights, and properties of the California Development Company and the two Mexican companies, in June 1916 (Dowd 1956).

Thus, IID, with its Mexicali subsidiary Compañía de Terrenos y Aguas de Baja California, secured the water for the Imperial Valley producers and, at the same time, the operation and management of the existing irrigation system in the Mexicali Valley. While this binational operating system existed, the Mexican company's managers were always Americans designated by IID. Even so, IID users were not satisfied because they did not have true control of the supply stream that flowed, for the most part, through Mexican territory. In addition, the almost certain expansion of agriculture in the Mexicali Valley, which would require ever greater volumes of water in the future, was yet another point of concern.

Frequently, the interpretation and judgments made of historical events, especially those that take place between two countries, contain elements of ethnocentrism—even more so in the case of historic events shared by border communities, as is the case of the development of the Mexicali and Imperial Valleys. This misinterpretation is usually applied to the role played by three companies involved in the development of Mexicali Valley's agriculture and economy. The first is the Colorado River Land Company, which had acquired most of the arable land in the Mexicali Valley and owned the agricultural irrigation canal system. The second was the Compañía de Terrenos y Aguas de Baja California (which took the place of the Sociedad de Irrigación y Terrenos de Baja California established in 1898) (Dowd 1956). This company was a subsidiary of IID and was in charge of operating and managing the water deliveries it needed.

The third company to play an important economic role was the Compañía Industrial Jabonera del Pacífico, which for many years mainly provided the necessary financing for cotton production. In addition to financing, the company provided (through a group of field inspectors paid by the company) technical support for the producers to ensure the best production for their clients. Most of the shucking and baling of cotton was conducted at the company's industrial plant in Mexicali. Soap was manufactured at the same site using cotton seed oil. The company had a compressor to extract cotton seed oil and obtained two important sub-products—the pulp left over from the extraction of cotton oil was used as cattle feed, and the cotton waste was used to fill mattresses and quilts. For several years the company also manufactured vegetable lard from cotton seed oil.

What criteria can be used for an objective interpretation of the role of these three companies? They were historically considered by Mexicans as enclaves of foreign interests and not particularly favorable to Mexican residents of the Mexicali Valley. As well, the American partners of IID ensured the prompt supply of the water they required. It cannot be denied that the Colorado River Land Company organized and rationalized, in some way, the exploitation of farmland in the Mexicali Valley. The fact that the policy was to rent or parcel out the lands preferentially to Chinese, Japanese, and Hindus instead of Mexicans made the company unpopular, which eventually led to its disappearance when its landholdings were expropriated under agrarian reform in Baja California.

The Colorado River Land Company marks the end of an historic time of land ownership in the Mexicali Valley. Given that the company owned significant parcels of land with agricultural potential in the Mexicali Valley, and that its business was to exploit these lands—which involved an ever-increasing demand for water—the possibility of growth in conjunction with the problems experienced before IID was created led to the conclusion that it would be necessary to divert Colorado River water to the Imperial Valley through a canal that was completely within U.S. territory (Hundley 2002).

The All-American Canal

Having achieved the organization of their irrigation district and, to a degree, having resolved the aforementioned problems, through IID the residents of the Imperial Valley began an intense lobbying campaign in Congress toward their goal of building the AAC, over which they would have absolute control. This canal would also free them of any intervention by or interference from Mexico that could jeopardize their water supply. Imperial Valley water users were concerned that their water supply, though under U.S. jurisdiction, immediately passed into Mexican territory—falling under the jurisdiction of the Mexican government—and that its flow was completely within the Mexicali Valley and its transportation was managed by a Mexican company that delivered the water to IID at the border. Note that the water delivered to IID not only covered the water demands of agriculture but also satisfied all domestic and industrial services for cities, towns, and farms (Dowd 1956).

In accordance with the 1904 concession by the Mexican government, Mexico had the right to half the water flowing through the canals in Mexicali, which to IID meant the loss of that amount in the available water supply for the Imperial Valley. With the AAC, this loss would be recovered and available for the expansion and development of the Imperial Valley. Likewise, there was the legal question about whether or not IID had the right to own a Mexican company and the possibility the Mexican government would take up the issue. An adverse decision could annul the concession by which the Imperial Valley water supply was allowed to be conveyed within Baja California territory in the first place (Dowd 1956).

Another problem facing the Imperial Valley residents was the price they could charge for the water delivered to Mexicali Valley users, which, by Mexican government decree, was held at a rate of 50 cents per acre-foot until 1919, which is when the Compañía de Terrenos y Aguas de Baja California announced that the authorized price was below actual cost. The Mexican government authorized an increase to 86 cents per acre-foot, which remained in place until 1931. The price of water charged to Mexicali Valley users was always a point of controversy between them and the landowners of the Imperial Valley, who argued that this quota was insufficient to cover the real operating costs and the reserves necessary to resolve flooding (Dowd 1956).

Making matters worse, in January 1931 the Mexican government authorized a 30% reduction in the price of water to Mexican users (the reduction was from 1.40 pesos per thousand cubic meters to 1 peso—in U.S. dollars it would mean a drop from 86 cents to 62 cents per acre-foot). The reduction was to be for just one year because the Mexican company planned to recoup the resulting losses in the following years. However, in December 1931, and without prior notification, the Mexican government issued a second order reducing the cost by an additional 25%. Upon protest from Compañía de Terrenos y Aguas de Baja California, this last order was rescinded; however, the amendment stipulated that the previous

30% reduction would remain in place indefinitely rather than for only one year. The Mexican government claimed poor economic conditions in the country made the cut necessary. In addition, Mexico abandoned the gold standard in July 1931, generating additional exchange rate losses for IID. This situation, coupled with a drop in farmable land, led to a loss of income from water sales to the Mexicali Valley. This income had averaged more than $550,000 in the previous eight years, but fell to $199,000 in 1931 and $86,000 in 1932 (Dowd 1956).

For many years the efforts of Imperial Valley residents on behalf of their hoped-for AAC failed due to the opposition of the legislators from Arizona, especially Carl Hayden. Starting in 1911 the attorney for the Imperial Valley, Phil Swing, and promoter Mark Rose helped the valley organize IID, which had sufficient legal standing to name directors, issue bonds, assess taxes, condemn property, and most importantly, buy and operate the valley's irrigation system. In 1917, IID managed to convince then-U.S. Secretary of the Interior Franklin Lane to order a study to determine the feasibility of the AAC. Secretary Lane agreed on condition that IID pay two-thirds of the cost. This condition was quickly accepted and the All-American Canal Board was formed, consisting of three engineers who, in December 1918, recommended construction of the canal at an estimated cost of $30 million. Strong opposition to the project came quickly from Harry Chandler, owner of the *Los Angeles Times*. Chandler headed a powerful syndicate of business owners who held more than 336,000 hectares (840,000 acres) in the Mexicali Valley that they rented to Mexicans, Chinese, and Japanese.

In the face of this threat, the response of the residents of the Imperial Valley was to "repudiate Chandler and associates and the 'Los Angeles Syndicate' in a referendum in January 1919 where the project was approved by an overwhelming vote" (August 1999). The opportunity for Imperial Valley residents to win their AAC presented itself in the Boulder Canyon Project, proposed by Arthur Davis, an engineer with the U.S. Reclamation Service, a "tireless defender of the concept of integral development of the Colorado River Basin" (Hundley 2002). "The term 'Boulder Canyon Project' hides a multidimensional task that had a profound effect upon the state, the West, the entire country, and northwest Mexico" (Hundley 2002).

Davis wanted to channel the efforts of Imperial Valley residents by convincing them that by joining the Boulder Canyon Project, which included projects to control the Colorado River through dams, they would win their canal more quickly. Davis would point out that the canal would not be safe against river surges without these control projects. Finally, his efforts worked and in April 1922 Swing, by now a congressman representing Imperial Valley, and Senator Hiram Johnson of California, formerly the state's progressive governor, introduced the Swing-Johnson or Boulder Canyon Bill, which provided for a large dam in the lower part of the river "at or near Boulder Canyon." It would be a hydroelectric power plant generating sufficient revenue to pay for the cost of the dam and the construction of the AAC (Hundley 2002).

Hayden continued to oppose the project with delay tactics and his political influence as one of the senior members of the House Committee on Irrigation of Arid Lands. He opposed the bill, saying it was "purely for the benefit of California" (August 1999). Finally, on May 25, 1928, the House approved the bill and sent it to the Senate, placing the Senate in the national spotlight.

> On May 28, 1928, now-Senator Carl Hayden participated in one of the most celebrated events in the history of the chamber; together with his colleague from Arizona, Senator Henry Ashurst, Hayden "filibustered" the Bill.[1] With only two days left in the session, the two Arizonians attempted to block the Bill by speaking... In his first speech from the floor of the Senate, Carl Hayden, "The Silent Senator," spoke without pause for nine hours before an almost empty Chamber. When Henry Ashurst took the floor, the opposing Senator, Hiram Johnson, proposed extending the session until June 5, in an effort to wear down the Arizonians; the motion was not approved, by a vote of 39-41. Accusations and threats of fisticuffs filled the Chamber while the Senate closed its session amid a "savage disorder" on May 29, 1928... An exhausted Carl Hayden confessed to a friend after his

> filibuster... "It is true that the Swing-Johnson Bill, if brought to a vote, would have been approved by both Houses... I sincerely doubt its passage can be avoided in the next Congress" (August 1999).

The 70th Congress opened its second session on December 5, 1928 and, in spite of Hayden's opposition, the Senate passed the Boulder Canyon Project on December 16 by a vote of 64–31. Two days later, the House passed the Senate's version 167–122. President Calvin Coolidge signed the bill three days later, putting an end to one phase of the Colorado River controversy (August 1999).

This important historical event was the beginning of an era of unparalleled development in the Colorado River Basin. The approved hydraulic projects included the construction of Hoover Dam and the AAC. "Few measures have had a greater impact on the Pacific Southwest than the action of Congress which authorized the construction of Hoover Dam and of the All American Canal in 1928" (August 1999).

The AAC ensured the water supply for the Imperial Valley at the expense of the growers in the Mexicali Valley. Upon the construction of the AAC it was no longer necessary to divert Colorado River water to the Imperial Valley through the Alamo Riverbed, which would surely affect the supply to the Mexicali Valley. An additional circumstance was that the irrigation system in the Mexicali Valley was owned by Compañía de Terrenos y Aguas de Baja California, a subsidiary of IID.

When the AAC began operating, Mexico began the fight for a treaty with the United States that would guarantee Mexican rights to Colorado River water. The 1944 Water Treaty assigns Mexico a guaranteed annual volume of 1.5 MAF of Colorado River water (Gobierno Estado de Baja California 1958).

The Boulder Canyon legislation, resulting in the construction of Hoover Dam which in its time, 1935, was one of the largest dams in the world, allowed the Colorado River to be controlled and therefore the possibility of building the All American Canal, which began to deliver water to the Imperial Valley in 1942. The water supplied was subsidized, as the federal government did not charge interest on the expenditures incurred in building the irrigation system, and the irrigable area of the Valley was stabilized at approximately 440,000 acres, providing in great measure a security and prosperity never before seen; it also helped to create and strengthen a landholding elite in the area (Hundley 2002).

> ...While the Boulder Canyon legislation wended its slow and troublesome way through Congress, the Imperial Valley was undergoing a great process of change over the entire acreage subject to irrigation, which was being transferred from small growers to a select group of "absentee landlords" with properties that averaged between 300 and 700 acres, and some as large as 3,000 acres (Hundley 2002).

A combination of good water years without the danger of flooding helped to generate millions of dollars of revenue for agriculture; these profits, the promise of a regulated river and the All American Canal had attracted large sums of investment capital from Los Angeles, San Francisco and other places. The financiers and speculators bought up the small parcels and transformed them into large-scale industrial farms with managers in charge who, naturally, sought to ensure the maximum yield. Most of the few growers who did not sell to the outside offers used their farming profits, expanded their bank credit sources to increase their properties, hired managers, and retired to the cooler temperatures and comforts of the coastal cities (Hundley 2002).

This left the Valley populated by a small handful of owners and operators at the top of the social pyramid and a great lower class of workers, mostly of Mexican origin, who worked for the agro-industry keeping the canals open and harvesting the abundant crops (Hundley 2002).

Three decades later, the average farm size was 200 hectares, with some as large as 2,000 hectares to 3,600 hectares (Hundley 2002).

> At the same time by then, the Imperial Valley was on the road to achieving one of the highest poverty levels in the nation, with close to 90% of agricultural workers (most Mexican by birth or paternity) classified as "lower class" in a study of social stratification in agricultural areas of

> the United States. One less point for the dream of small irrigated farms and social reform in the Imperial Valley (Hundley 2002).

Frequently, history provides important lessons about how the aspirations of societies and men of vision are achieved. History also provides interesting contrasts. The development processes of the Mexicali and Imperial Valleys, as well as the results manifest in each today, are one of these lessons in contrast. While in the Imperial Valley farmland came to be concentrated in the hands of a few, in the Mexicali Valley agrarian reform *(reforma agraria)* in 1937 distributed the extensive landholdings of the Colorado River Land Company to *ejidos* of Mexican farmers who did not own land.

Agro-industry in the Imperial Valley generates more than $1 billion annually. However, Imperial County is one of the most economically depressed counties in California, with an unemployment rate near 30% and an economy based on agro-industry, commerce, and services. According to the last census its population is 140,000. A large percentage of the labor force used in commerce, services, and agro-industry in Imperial County reside in the city of Mexicali and cross the border daily to work and to return home. The county does not have a diversified industrial sector. The economy of the City of Calexico, based mainly on commerce and services, depends completely on retail purchases made by the population of Mexicali.

The Mexicali Valley, though, currently does have a diversified agro-industry, which still provides important income for the region. An important industrial sector is based in 21 industrial parks. A strong home-building industry enjoys constant demand due to constant demographic growth. And Mexicali has large and modern shopping centers. The total population of the municipality of Mexicali is 764,602, and the city of Mexicali has a population of 549,873; the unemployment rate is 10% (INEGI 2000). There are two important universities in Mexicali, including Universidad Autónoma de Baja California, which as a public institution is one of the most highly regarded universities in all of Mexico. One could conclude that the "American dream" of Oliver Wozencraft, Charles Robinson Rockwood, and W. T. Heffernan has been more than achieved in Mexicali, but not in the Imperial Valley.

CONCLUSIONS

This chapter described the historical development of an extraordinary battle for control and use of Colorado River water that began in 1849, and in which the main protagonists were, and continue to be, the residents of the vast southwest region of the United States, home of the Colorado River Basin and an area where extensions and diversions of the river's waters have benefited these residents. Despite the fact that, at certain historical moments, parts or elements of the great controversy are resolved by reaching a consensus, conflicts continue today in the disposition of certain types of waters, such as the AAC, where elements of international relations are also involved. But what is interesting for these conclusions is trying to establish some of the features of "hydropolicy" and the water culture generated by the experiences and actions of those responsible for carrying on the fight, those who managed to negotiate the necessary agreements to achieve an equitable distribution of Colorado River water between the seven southwestern U.S. states that share its basin. It is possible that the mentality of those currently responsible for water policy in the United States, and especially in the state of California, corresponds to the original perspectives of their predecessors. If that is the case, Mexican negotiators must keep this background in mind when designing their negotiating strategies. But they must keep in mind, too, the level of new, shared interests between Mexico and the United States regarding North America, especially as represented by the North American Free Trade Agreement (NAFTA).

Several aspects of water policy, or "hydropolicy," became criteria of general application. The are detailed below.

The Concept of Water Property in the United States

- The concept of "first come, first served" was used in California to resolve controversies over who had water rights, which were always scarce in California, between gold prospectors during the "Gold Rush"

- States, based on their autonomy, have the right to determine the use of the natural resources in their boundaries
- Water users own the resource and can freely dispose of or reclaim at any time the use of water over which they have primary rights and over which they may have lost or ceded temporary control

A recent example of these above points is the fact that for many years California used part of the water quota assigned in the Colorado River Compact to the state of Arizona, which could not use the entire assigned volume. Currently, and due to the significant accelerated development of the Tucson and Phoenix areas, Arizona notified California that it may not continue to use this water. This is the same criteria maintained by IID with regard to the groundwater problem generated by seepage from the AAC.

The Concept of Waterworks and Land Reclamation Programs

The federal government, with federal funds, is the only entity capable of undertaking the necessary irrigation and hydroelectric projects.

The Concept of Irrigation District and Systems Management

The federal government, once the project is concluded and if it is public, must turn it over to local control for its care, maintenance, management, and control. It is important to distinguish between the term "hydropolitics" used by August (1999), and the concept of "water culture." Hydropolitics, for August, has the connotation of an historically long legislative process that determined public policy for the use and management of the resource, assigning clear roles to the state and to the various citizen, water-using actors. Water culture refers unmistakably to an historically long social process that begins with the arrival of the first pioneers to colonize the U.S. Southwest. They developed a perception about water—a scarce resource—and very early on identified what access to it would mean for the welfare, progress, and security of their communities. The concept of "first come, first served" was the result of these settlers' experiences. From the beginning, the settlers determined that the use of water would be primarily for agriculture, and the defense of their right to that use determined their demands in terms of legislation that would ensure that objective.

Endnotes

[1] "Filibuster" (obstruction) is a U.S. political strategy used by legislators to prevent a vote on a bill or issue under discussion. The member requests the floor and speaks non-stop, as long as necessary for the session to end and the matter to be held over for voting in the next session. The member may not stop speaking or yield the floor to a member of the opposing party. Members of the same party may take turns speaking during the filibuster.

References

August Jr., J. L. 1999. *Visions in the Desert: Carl Hayden and Hydropolitics in the American Southwest*. Fort Worth, Tex.: Texas Christian University Press.

Dowd, M. J. 1956. *History of Imperial Irrigation District and the Development of Imperial Valley*. El Centro, Calif.: Imperial Irrigation District Community and Special Services.

Gobierno del Estado de Baja California. 1958. *Memoria del Primer Congreso de Historia Regional Volumes I and II*. Mexicali: Gobierno del Estado de Baja California.

Heffernan, W. T. 1930. *Personal Recollections*. Calexico, Calif: Calexico Chronicle.

Hundley Jr., N. 2002. *The Great Thirst. Californians and Water: A History*. Berkeley, Calif.: University of California Press.

Instituto Nacional de Estadística Geografía e Informática. 2000. *IX Censo General de Población y Vivienda 2000*. México, D.F: INEGI.

Medina Robles, F. 1970. "Mexicali-Caléxico, Estudio Comparativo de su Desarrollo." Professional thesis. Universidad Autónoma de Baja California, Mexicali, B.C.

Rockwood, C. R. 1930. *Born of the Desert*. Calexico, Calif.: Calexico Chronicle.

VI

Looking Across the Canal: Reflections on Visions and Policies on Water Issues in the United States

María Rosa García Acevedo

Introduction

The All-American Canal (AAC), located in California's Imperial Valley and running parallel to the U.S.-Mexican border, recently attracted attention in the United States, having acquired the sad notoriety of a "high-risk" area for Mexicans emigrating into the United States (Bureau of Citizenship and Immigration Services 2001). A Border Patrol spokesman in Calexico, explaining why his agents were once found on Mexican soil, acknowledged there is no divider in the area where they were found and noted the border markers are several miles apart. The situation, in his opinion, is such that many Mexicans—and even Border Patrol officers themselves—sometimes confuse the AAC for the international limit line between the two countries (*The Washington Post* 2000).

But the AAC has been part of the U.S. domestic policy agenda, too. The U.S. government intends to have the currently porous canal lined with concrete by December 31, 2008; it claims this will prevent water "waste"—caused by seepage into the aquifer—of 67,700 acre-feet per year (AF/y) (Cortez Lara and García Acevedo 2000). This approach ignores the fact that, in terms of water flow, the Imperial and Mexicali Valleys are part of the same geographic area. The history of bilateral water arrangements has received little attention, as well. The problem is that, as mentioned in several chapters in this volume, lining the AAC will stop the flows that have historically fed the wells in the Mexicali Valley. This lining will have negative consequences on agriculture, the environment, and the overall quality of life of area residents.

In this chapter, the AAC situation will serve as a window through which to examine the different visions of water that have ruled in the United States since the end of the 19th century and the policies these visions have helped shape. This chapter supports the Blatter, Ingram, and Levesque thesis that the vision of and discourse on the development of the different meanings of water in the United States have inspired a broad variety of public policies, fostered opportunities, set limitations on the actions of participating political stakeholders, and created winners and losers on both sides of the border (Blatter, Ingram, and Levesque 2001). What also stands out is how the classic vision of a nation-state, where U.S. public policy would be designed exclusively for those living within its borders and would only have consequences within the country, has not always been applied to transboundary water resources, and in particular to the AAC. The specific questions addressed have to do with the effect of the current U.S. perception of water as a commodity, particularly water in the Lower Colorado Basin from which the AAC is supplied. The consequences of lining the AAC on the future water-flow permeability of the U.S.-Mexican border are also explored.

This chapter is divided into three parts. The first reviews the different visions of water and how they have affected policies for the Colorado River since the second half of the 19th century. Through these visions the origins and management changes of the AAC can be traced, and thus the evolving view of the political boundary separating the two countries. Water was originally seen as part of nature and attached to its natural environment. Later, water was seen as a product that engineering could convey from one place to another. At times, water was perceived as a security tool, critical to national survival and requiring the attention of policymakers at the highest level of government. The second part of the chapter focuses on the current water markets and how U.S. policymakers have interpreted issues related to the management of the Colorado River and its transboundary implications. From this point of view, water would be subject to market conditions, which would define its movement from one place to another and from one use to another (Blatter, Ingram, and Levesque 2001). New projects were implemented under this vision, including the AAC lining proposal. The third section closes with the current possibilities for and limitations to discussing the transboundary flows of Colorado River water through the lens of the AAC case study.

Historical Context: From Planning to Operation of the Canal

When the water in the Imperial and Mexicali Valleys flowed in its natural environment, no need existed for canals to modify its course. The native populations, such as the Cucapa, moved from one place to another according to the agricultural cycle, following what the Colorado River dictated. Their culture revolved, literally and figuratively, around water. Even after the international border between the United States and Mexico had already been established as it exists today, these indigenous populations still moved freely across it (Álvarez de Williams 1975). The United States and Mexican governments played no part whatsoever in this movement.

Toward the end of the 19th century the mobility of the native populations, which had been associated with the free flow of Colorado River water, was fundamentally changed. A privately financed project for manipulating water by using a canal to transport it elsewhere was put in place—without taking into account the interests of this water's original users. The first American and Mexican explorers of the Imperial and Mexicali Valleys were inspired by this vision of water. In 1896, investor C. R. Rockwood proposed conveying the waters of the Colorado River all the way to the Imperial Valley, with the intent of promoting economic development. To that end, the same year he founded the California Development Company (Metz 1989).

The idea of permeable borders goes hand in hand with the characterization of water as a product. In that sense, as a possible solution to the technical problems associated with building a canal on

U.S. soil (specifically the existence of a sandy plateau), the California Development Company opted to convey the waters of the Colorado by means of a binational canal. The argument was that it was cheaper and more feasible, due to the technology available at the time, to use the Alamo Canal, which had a 45-kilometer (km) (28-mile) stretch going through Mexico (in the Mexicali Valley), to irrigate the Imperial Valley (Fradkin 1981).

To create a binational endeavor and overcome the restrictions set forth in Mexico's laws, California Development Company and its successor, the Colorado River Land Company, supported the formation of the Sociedad de Irrigación y Terrenos de Baja California (in English, Baja California Irrigation and Land Company). Guillermo Andrade, a businessman and former Consul of Mexico in Los Angeles, headed the Sociedad de Irrigación. In the end, this agreement between private companies became the only legal framework for using the Alamo Canal and supplying an annual allocation of 3.63 million acre-feet (MAF) (4.478 billion cubic meters [m^3]) of water from the Colorado River to the Mexicali Valley. This arrangement between non-government stakeholders—individuals and private companies—with the tacit approval of the governments, brought forth the possibility of agricultural development in the Imperial and Mexicali Valleys (Metz 1989).

During the first decade of the 20th century, the vision of water as a product overlapped with the vision of water as a security tool. The change took place as a result of the 1905–1907 floods that affected both the Imperial and Mexicali Valleys. An engineering failure was to blame for the disaster. These floods diverted water from the Colorado River to the Salton Depression, "artificially" creating the Salton Sea. For the first time, the U.S. federal government, and even President Theodore Roosevelt himself, considered the water of this area essential to national security and became involved in the flood issue. Roosevelt declared a state of emergency and granted the Southern Pacific Railroad Company $3 million to wage a "battle against the Colorado" (Gottlieb and Fitzsimmons 1991).

Because water was perceived as essential to national survival in the United States, the crisis caused by the floods generated a wave of criticism against the permeability of the U.S.-Mexican border with regard to water flows. This marked the first time the possibility of using a canal built exclusively on U.S. soil, instead of the Alamo Canal through Mexican territory, was ever discussed. It was the association of Imperial Valley agricultural users who initially promoted this project. The project later won the support of the U.S. Department of the Interior's (DOI) Bureau of Reclamation (BOR), the Imperial Irrigation District (IID), and various members of the U.S. Congress. The greatest obstacle then became the high cost of a new canal project (Fradkin 1981).

The topic of building a canal on U.S. soil remained on the American agenda during the 1920s. In 1922, Arizona, California, Colorado, Nevada, New Mexico, Utah, and Wyoming signed the "Colorado River Compact." Citing political instability, Mexico was not invited to take part in the negotiations (Fradkin 1981). Per the provisions of this agreement, the Colorado River Basin was divided into upper and lower basins and water for each state was formally apportioned (Postel, Gleick, and Morrison 1996).

Finally, the plan to build a canal on American soil to convey Colorado River water to the Imperial Valley was set in motion by the approval of the Boulder Canyon Project Act of 1928 (Fradkin 1981). Successful lobbying in favor of the canal, together with the development of technology that would make this infrastructure project possible, led to its approval. The decision to build the new canal was linked, under the Boulder Canyon Project Act, to other important provisions on Colorado River water, including the construction of Hoover Dam and its adjacent Lake Mead. This act also established California's right to receive 4.4 MAF/y of Colorado River water. In Mexico, the act marked the beginning of the artificial regulation of the flow of the Colorado River, which at times had serious supply consequences (Henderson 1968, Ward 1999, Stapleton 2001).

The Great Depression, as well as other domestic factors in the United States, delayed construction of the AAC, and therefore it was not until 1942 that it was finished and began operating. The canal was given the symbolic name "All-American" to underscore the victory its construction represented to the United States in matters of security (Fradkin 1981). Since then, the canal has irrigated 1 million acres in Southern California and has been key in the development of the Imperial Valley, whose irrigation district (the IID) is the largest user not only in California, but in the entire Colorado River Basin system (Stapleton 2001).

During the 1940s, the vision of water as a security tool resurfaced in the United States in another context—that of cultivating a close relationship with the countries of the Western hemisphere (including Mexico), in light of World War II. This began the first formal talks between the U.S. and Mexican governments on their border-region rivers, the Río Grande (Río Bravo in Mexico) and the Colorado River. These negotiations led to the signing of the Treaty for Utilization of Waters of the Colorado and Tijuana Rivers and of the Rio Grande in 1944, referred to as the 1944 Water Treaty. The U.S. president played a crucial role in ensuring this result. In supporting this treaty, Franklin D. Roosevelt offered guarantees to the Mexicans that their concerns would be heard along with those voiced by the Colorado River states. His support of the treaty was a decisive factor for its approval by Congress. Under this treaty, Mexico was assigned 1.5 MAF/y (the quality of water to be received was never discussed). It also stipulated that the International Boundary Commission, a bilateral institution founded in 1899, would become the International Boundary and Water Commission (IBWC), to be composed of a Mexican and a U.S. section. This commission was put in charge of all matters related to crossborder water resources (Metz 1989; Sánchez Ramírez 1990).

Years later, in 1973, the view of water as a security tool was successfully championed by Mexico before the United States, and this led to the amendment of the 1944 Water Treaty (Fradkin 1981). At the core of the new agreement, IBWC Minute 242 centered on solving the issue of Colorado River salinity, which had ravaged Mexicali Valley agriculture for more than a decade. Minute 242 also brought the issue of transboundary groundwater to the arena of the treaty. Both governments made a commitment to inform the other prior to any project involving groundwater, as the case of lining the AAC would later be. Moreover, Minute 242 also left open the possibility of coming to an agreement on the matter (Utton 1991, Mumme 2000).

The different visions of water that have dominated in the United States since the second part of the 19th century have inspired a broad variety of policies on Colorado River water. The planning, construction, and operation of the AAC have been a significant part of them. Undoubtedly there have been opportunities, but there have also been limitations in the recognition of the transboundary nature of Colorado River water. The stakeholders involved have been from both government and the private sector. The latter, at times, played the most salient role.

The Era of Water Markets and the Canal-Lining Project

In the United States, as in many other parts of the world, the argument that water markets can manage this vital resource more efficiently has gained strength (DiMento 2001). In the debate over Colorado River water, several topics have arisen and become intertwined, including rural-urban water transfers, conservation projects, compliance with environmental regulations, and the new arrangements among U.S. users of the Colorado River. The question of how the AAC has been immersed in the discourse and policies associated with these important topics will be discussed in this section.

Rural-Urban Water Transfers

The Imperial Valley, where the AAC is located, has become an ideal region for the transfer of water from rural to urban areas in Southern California because IID receives the largest amount of water from the Colorado River Basin (at a current rate of 13.85 MAF/y), mainly for agricultural uses (Brazil 2001). As well, the Imperial Valley is within the proximity of Southern California cities such as Los Angeles and San Diego, which are eager to be part of a stable water market, one that would allow them to meet the increasing needs of their growing populations (Rohrlich 1998). San Diego is particularly dependent on Colorado River water because it has virtually no groundwater or other supply sources (Perry 2003a).

It is a misconception that the sale of water does no harm; this notion ignores the fact that water sales create disparity. Their eventual negative impacts on the United States have not been discussed in-depth, much less have their crossborder impacts (Perry, Blatter, and Ingram 2001). For example, a water transfer will cause a social polarization in the Imperial Valley, where only the large landowners will profit from water transfers and the interests of small- and medium-sized agricultural producers will be ignored (Blatter, Ingram, and Levesque 2001; Gottlieb 1988; Perry 2003b).

Beginning in the 1980s, several U.S. government agencies actively became involved in the discussion of how water transfers from the Imperial Valley to urban areas in Southern California should be conducted. A prominent player has been the U.S. Secretary of the Interior, who has used the powers granted by the 1963 U.S. Supreme Court decision in *Arizona v. California*, which upheld the 4.4 MAF/y allocated to California under the Boulder Canyon Project Act of 1928 and granted Arizona 2.8 MAF/y and Nevada 300,000 AF/y. According to this ruling, the Secretary of the Interior has the legal mandate to determine, on an annual basis, the existence of a surplus, normal conditions, or drought in the Lower Colorado River Basin (Stapleton 2001, Sierra Club 2002). A year later, the Supreme Court authorized California to use surplus water taken from Lake Mead.

Other key participants in the water transfers have been the public agencies in charge of water management in Southern California, such as the Los Angeles-based Metropolitan Water District (MWD), serving more than 17 million people through 26 local agencies; its subsidiary, the San Diego County Water Authority (SDCWA), serving nearly 3 million people; the Coachella Valley Water Authority (CVWA), serving the area adjacent to the Imperial Valley; and IID, the largest Colorado River water user and the location of the AAC.

Amidst difficult negotiations—labeled as water-control "wars"—and with the full support of DOI, SDCWA and IID signed an important agreement in April 1998 (SDCWA 1998a). Under the agreement, IID was to supply 200,000 AF/y of water to San Diego County for 75 years beginning in December 2002 (SDCWA 2001). DOI portrayed this agreement as an example of how water transactions brought nothing but positive results for all parties involved, as well as guaranteed supply stability. As will be discussed later, this agreement also became a key component of new arrangements for

Colorado River water (specifically the Quantification Settlement Agreement, signed by Southern California water agencies) (Perry 1997, SDCWA 1997).

The aforementioned water-transfer agreement between Imperial Valley and San Diego was not put into place in December 2002 because of the complexity of its implementation, including the need to overcome obstacles regarding environmental protection of the area. The delay in the water transfer agreement affected the start date for the AAC lining project; it was not until October 2003 that all the pieces of the puzzle fell into place. In the face of intense negotiations among Southern California water agencies, the action of the federal and state governments was a decisive factor in overcoming the obstacles for water transfers, to the point that, according to some observers, this was not a water-sale transaction dictated by market forces but an exchange performed under intense pressure from the U.S. government (Leavenworth 2003).

Water Conservation Projects

In the framework of water transfers, the United States has promoted water conservation policies that by themselves probably would not have had enough momentum to move forward. California, in particular, linked water conservation projects to rural areas under the assumption that water is wasted in agriculture. The Imperial Valley, because of the volumes of water it brings in, became an ideal place to implement water conservation policies.

U.S. policymakers presented the canal-lining projects in the Imperial Valley area, including the AAC, as necessary support for precisely these conservation efforts. In 1998, the U.S. House of Representatives held hearings about lining the AAC. The topic was presented as a simple "water-saving" effort that would allow California to reduce its consumption in the long term (to 4.4 MAF/y) and bring nothing but benefits to everyone. IID voiced its opposition from the outset. The agency feared that greater involvement by the federal government in the control of the AAC would limit the role of IID in the long run. This agency even suggested other means of increasing water savings, such as the lining of secondary canals (Public Law 100-675, House 73-81). In fact, during the 1980s, IID had received nearly $200 million from MWD for conservation projects, with the stipulation that MWD would obtain whatever water was conserved (McClurg 1996, Rohrlich 1998, Sierra Club 2002).

In the end, with an astounding lack of technical data, the 1988 law authorized the Secretary of the Interior to reduce water seepage from the AAC to the aquifer by means of lining portions the canal (Mumme 2000). The lining project could not begin, though, because the 1998 law did not allocate money for that purpose. In the meantime, in 1994, BOR studied several options for the AAC, including building a parallel lined canal, drilling wells near the canal to recover water that had seeped through, lining 25 miles of the existing canal, and taking no action at all (Jones, Duncan, and Mumme 1997).

There were other efforts in the 1990s to move forward with the lining. In 1996, for example, MWD and the Southern Nevada Water Authority planned to line the canal, take the water "saved" through this process to Lake Mead, and later divide it equally between themselves. While the project, called Reliability Plus, never materialized, it does show how the argument for lining the AAC was used again in the context of conserving water in rural areas to increase urban-supply availability (*Los Angeles Daily News* 1996).

Funding for the lining project was finally obtained in September 1998. Under the framework of the conservation debate, the lining project reached the California Senate agenda (Rohrlich 1998). Senator Steven Peace (D-El Cajon) managed to negotiate the final agreement to fund the project by attaching it to an endangered forest area issue supported by Northern California legislators. The California Senate approved the measure by a 33–3 majority vote (SDCWA 1998b, Gardner 2001), appropriating $200 million for the lining of portions of the AAC by December 2006, barring "extraordinary circumstances." The same legislation appropriated $35 million for necessary infrastructure to store the groundwater recovered as a result of the lining (California Senate 1998). MWD strongly supported this proposal, given that more water would flow into its distribution system. At the federal level, DOI also highlighted the advantages of the project (McKinnon 2000).

The approval of this 1998 legislation was a huge step for California's water market endeavors. In November 1998, the particulars of the pricing of Imperial Valley-San Diego water transfers were approved (Associated Press 1998). However, it was not until October 2003 that the final details of the historic IID-SDCWA water transfer agreement were finalized, including issues related to environmental conservation in the Salton Sea area.

Environmental Provisions

Environmental conservation topics have been part of U.S. legislation since the 1970s. In the case of the AAC, from lining approval under the 1988 legislation to date, there have always been environmental legislation provisions with which to comply. Specifically, Public Law 100-675 states that the Secretary of the Interior must submit a report to the U.S. Congress on the project's effects on wildlife in the region (Public Law 100-765; La Rue 2001a). As noted by Mumme and Lybecker in their chapter, an environmental impact study was submitted in 1994.

AAC-related environmental issues were later linked to others in the same geographical area, particularly to those of the Salton Sea. SB 1765, approved by the California Senate in 1998, allocated $300,000 to the Salton Sea Authority (created in 1993) for new studies on AAC seepage and surface flows to determine how much water would be lost to canal lining and the environmental impact of decreased inflow (once the canal were lined) to the Salton Sea on the species that inhabit the surrounding area. The authority was also given a mandate to formulate a plan to mitigate the effects of those environmental impacts (California Senate 1998).

In fact, compliance with environmental regulations in the Salton Sea area was a crucial element that stalled the IID-SDCWA water transfer agreement. Opposition focused on the argument that the water transfers would reduce the Salton Sea inflow and thus negatively impact area species (Brazil 2001). In addition, IID did not want to become legally liable for the environmental impact on these species if the water transfer to San Diego began, per the agreement signed in 1998.

Toward that concern, DOI and the state government of California put pressure on California public water agencies to expedite the water transfer process. The problem was that California risked losing its right to use Colorado River surplus waters should the state fail to finalize the transfer of water agreement by December 2002 (Brazil 2001). To that end, several bills were introduced (Raley 2001). On a federal level, for example, HR 2764 in August 2001 allocated funding for a habitat conservation program to protect endangered species in the Salton Sea area (U.S. House of Representatives 2001). In addition, Public Law 100-675 provided funding for infrastructure and reservoir projects in the area surrounding the AAC that would mitigate the effects of reduced water flow on the region's habitat. Its aim was to avoid a substantial decrease in Salton Sea inflows as a consequence of lining the AAC (U.S. Senate 1988).

On the state level, in September 2003 California approved SB 654, which authorized the taking of species incidental to facilitating the water transfer (California Senate 2003c). Also, California legislators passed bills SB 277 and SB 317, appropriating $300 million in funding to restore Salton Sea habitat (California Senate 2003a, California Senate 2003b). These actions contributed to finalizing the IID-SDCWA water transfer and expediting the announcement that SDCWA would be in charge of lining the AAC.

New Arrangements among U.S. Colorado River Users

In the 1990s, while the debates about water transfers, conservation, and environmental protection were taking place, disputes arose among U.S. states along the Colorado River. Arizona and Nevada were reaching their 1963-assigned water allocation levels and sought increases to meet the needs of their growing populations (Sierra Club 2002). The precarious balance reached after the 1963 Supreme Court decision was about to come undone. As a solution to the challenge of meeting their new water needs, Arizona and Nevada pressured California to decrease its use of water from the Colorado and limit its use of surplus water from this source. In fact, surplus waters gave California a *de facto* increase in its allocation. California uses as much as 5.5 MAF/y, surpassing its 4.4-MAF/y allocation (Sierra Club 2001). In addition, Arizona and Nevada asked DOI to establish specific rules for the use of Colorado River surplus water, which until then had been used only by California (Snedeker 2000a).

Amidst difficult negotiations in which DOI fully exercised its legal capacity to distribute flows from the Colorado River, California was persuaded to limit its long-term water consumption from this source to 4.4 MAF/y. In exchange, California received a 15-year grace period (until 2016) during which it would have access to surplus water. In this scenario, Southern California water agencies, including IID, MWD, and SDCWA, negotiated new rules amongst themselves for the distribution of water. In May 2000, California presented Plan 4.4, the Colorado River Water Use Plan, a complex document reflecting the consensus reached by all these agencies on the protection and optimization of water resources from the Colorado River. Both the implementation of the Imperial Valley-San Diego water transfer agreement and lining of the AAC were key parts of this plan (U.S. House of Representatives 2001, Stapleton 2001).

Plan 4.4 was a key prerequisite for the signing of a broader agreement by the riparian states of the Colorado River Basin. In January 2001, Arizona, California, Colorado, Nevada, New Mexico, Utah, and Wyoming signed the Record of Decision for Colorado River Interim Surplus Criteria (Raley 2001, Walters 2000). This agreement acknowledged California's effort to limit its water use, ratified the commitments made in Plan 4.4 before the other states, and included a schedule of specific measures (including the water transfers from IID to SDCWA) that would be taken to reduce California's water use starting in December 2002 and running through 2016 (Brazil 2001, Stapleton 2001). By adhering to this time table, California would be allowed to use the surplus water, which in 2002 amounted to 600,000 AF/y (Brazil 2001, Ritter 2001).

Lining the AAC was also a key piece of these interstate negotiations. According to sources from Arizona, California's "rival" in the negotiations, California proffered the argument for continuing to use the Colorado River water surplus (stored in Lake Mead) in exchange for its commitment to line the AAC, thus substantially moving water conservation policy forward (Pearson 1996). Moreover, in September 2003, once the implementation of the Quantification Settlement Agreement was formalized, MWD, SDCWA, and CVWA reached the historic agreement to begin the "largest and longest-lasting water transfer in history," according to Jimenez (2003). Under these conditions, SDCWA assumed the commitment to line the AAC, at a cost of $327 million ($235 million from California state coffers, which had been authorized in 1998, plus $90 million to be funded by SDCWA), stipulating that the "saved" water would flow to SDCWA. The agreement also postponed the deadline to line the AAC to December 2008 (Jimenez 2003, SDCWA 2003).

The era of water markets has brought many changes to the Lower Colorado River Basin, including the new arrangements among California public water agencies for water use and conservation and the Record of Decision for Colorado River Interim Surplus Criteria agreement between the riparian states of the Colorado River. Thus far, although negotiations have been difficult, agreements have indeed been reached. Lining the AAC became the cornerstone of these negotiations because it was attached to the most ambitious rural-urban water transfer in the history of the United States (signed between IID and SDCWA). In fact, the only limitation—the compliance with legal provisions regarding environmental protection—was overcome through the intervention of California policymakers that found a way to reduce environmental restrictions and grant funding to environmental restoration projects (Perry 2003b).

The issue of lining the AAC went beyond being of mere "local" interest or an isolated issue in the U.S. public policy agenda to a cornerstone of 21st century U.S. water policy. Such a turn of events certainly makes it more difficult to avoid the implementation of the lining project. However, questions remain as to whether there is room in this era of water markets for a debate about the transboundary consequences of the new arrangements related to the Colorado River, and more importantly, the possibility of introducing into the discussion on the AAC other visions of water more favorable to the idea of a permeable border.

The Lining of the Canal and the Issue of Border Permeability

In the U.S. debate on the lining of the AAC, there has been little consideration of its repercussions in Mexico. The U.S. obligation to formally consult Mexico on groundwater issues, as legally set forth in Minute 242, has been limited to "informing" Mexico, mainly through the Mexican section of IBWC. For example, when the first hearings on the issue of the AAC were held in the U.S. House of Representatives in 1998, DOI and BOR stated that the harm to Mexico, if the lining project were to take place, would be "insignificant." Later, when U.S. agencies (specifically IID) finally acknowledged the potential damage to agriculture in the Mexicali Valley as a consequence of lining the AAC, they never recommended holding formal conversations with Mexico (U.S. House of Representatives 1988).

During the discussions of HR 2764, the effects on Mexico of lining the AAC were only superficially included. This bill authorized DOI to "increase Mexico's ability to more effectively use its water allocation under the Treaty of 1944." Also, it instructed DOI to prepare a report in December 2003 about consultation with Mexico on any water management issues. Finally, the text of this bill proposed coordination between DOI and the U.S. commissioner of IBWC on possible consultation with Mexico, leaving the possibility open that public water agencies in California (MWD and SDCWA) might participate. However, the justification for these actions was presented as a "courtesy" to Mexico and not as an acknowledgement of the legal requirement for such consultation (U.S. House of Representatives 2001).

Another obstacle to holding fruitful negotiations with Mexico on the AAC lining project was the fact that IBWC has not historically shown particular interest in groundwater issues (Jones, Duncan, and Mumme 1997). Yet, IBWC has taken some token actions. In May 2000, in response to Mexican border region senators, IBWC officials raised the possibility of asking the United States to compensate Mexico for the water losses caused by the lining project. They even proposed alternatives, such as having Mexico receive a larger water allocation from the Colorado River by having the AAC deliver 500 more cubic feet per second to the Mexicali Valley (Hume 2000, Restrepo 2000). IBWC also studied other means of mitigating the impacts of lining the AAC through water conservation programs in Mexico (Dibble 2004).

However, when the California state laws in 1998 and 2003 authorized funding and gave the green light for lining the AAC, there was no longer any mention of possible repercussions on the other side of the border. This could lead to the conclusion that in the water market era, water-flow permeability across the border no longer exists. Yet, in contrast, the United States and Mexico have celebrated formal consultations on the possible construction of a new binational aqueduct that would provide potable water distribution services to the metropolitan areas of San Diego and Tijuana-Rosarito. This canal would be built parallel to the Colorado River Aqueduct and would run through Mexican territory

It has been precisely SDCWA, the same agency in charge of lining the AAC, that has expressed interest in exploring the crossborder aqueduct project. In fact, SDCWA funded a study that focuses on the feasibility of this project. SDCWA's interest is twofold: First, it would allow this agency to become almost completely independent from MWD because it would stop using the latter's aqueduct to convey "its water" southbound through California. Also, SDCWA would have an estimated cost-savings of $2 billion if this aqueduct were built in Mexico instead of in the United States (Conaughton 2001, LaRue 2001b). SDCWA's interest in a crossborder project suggests that even in this era, and when it fits their interests, some U.S. water agencies may consider negotiating on issues that go beyond their own and their country's borders. So then, the issue of lining the AAC, or at least of its modalities, could be subject to negotiation in a broader spectrum that included other transboundary issues, such as the binational aqueduct. SDCWA, in any case, could become an interesting counterpart in negotiations with Mexico.

On the other hand, the issue of lining the AAC could be examined through other visions of water that overlap with the dominant one of water markets. This could lead to different scenarios—water could be considered a security tool and a tool to create bonds between communities. The federal government in Mexico (during the administrations of Alvaro Obregon, Manuel Avila Camacho, and Luis Echeverria) considered the waters of the Colorado River a matter of security that should be included in the bilateral agenda between the two countries. Getting the attention of their U.S. counterpart proved difficult, but historical precedents, such as the signing of the 1944 Water Treaty and Minute 242, suggest that success can be achieved.

A recent example of statements about water as a security tool took place during a meeting between Mexican president Vicente Fox and U.S. president George W. Bush in February 2001. Fox proposed comprehensive negotiation on transboundary waters. He declared that "we are working on the whole issue of water in the border region, both rivers, the Colorado and the Río Grande, and we are working on them together" (Office of the Presidency of the Republic 2001). However, Mexico's decision to deal with Mexico's Rio Grande "water debt" in 2004 as a separate issue was a step backward in the potential joint high-level negotiations regarding both border-region rivers.

Another scenario would be related to envisioning water as a tool for creating bonds between Mexican communities and perhaps binational bonds as well. Using this lens, water would be linked to regional development and a true preservation of the habitat, placing these values above those related to water markets. In this case, as Alfonso Cortez Lara points out in his chapter, the main stakeholders would be Mexican non-governmental organizations in possible partnership with their counterparts in the United States. The Sierra Club, for example, has called the AAC lining project a "simple plan" that will exacerbate other problems in the area. The Sierra Club focuses on the problems water loss would cause in the Imperial Valley, the fact that the environmental impact assessments stop at the border, and also acknowledges that Mexico could lose more water than what the United States has considered thus far (Sierra Club 2001 and 2002).

The seeds of this view of water also exist in Mexico. Its historical precedent is a grassroots movement that made the Mexican federal government recognize the seriousness of the Colorado River salinity problem in the 1970s. Today, according to data collected by a COLEF-Mexicali survey in 1998, 67% of Mexicali Valley water users are aware of the damage that lining the AAC will cause them in terms of income, crops, and production costs (Cortez Lara and García Acevedo 2000). Although protests against the lining project have not been as visible as in the past, one cannot rule out that this might change, particularly once its effects are felt in the Mexicali Valley. In December 2004, for example, Mexicali Valley farmers, along with state legislators and business group representatives, raised their voices against the "submissive attitude" of Mexican federal authorities toward the United States on AAC-related matters (Dibble 2004).

An example of efforts to create binational bonds between U.S. and Mexican environmental groups over the Colorado River is the Santa Clara Marsh. Successful social mobilizations were taken as far as Washington, D.C. Their impact was reflected in the Record of Decision for Colorado River Interim Surplus Criteria agreement. The agreement notes the need for special measures to protect the Santa Clara Marsh and opens the door for subsequent negotiations to grant more water to Mexico to preserve the habitat of the Colorado River Delta (Snedeker 2000a and 2000b).

Looking Across the Canal: Final Thoughts

The All-American Canal is sometimes confused with the U.S.-Mexican border. This is not mere coincidence—its geographical location and history are woven together with binational events and policies. However, with regard to water flow, there have been times when the AAC has been seen in the United States as exclusively a domestic policy issue. Nowadays, for example, its fate is intimately linked to the "largest and longest-lasting water transfer agreement in history" (Stapleton 2001). Moreover, as a result of lining the AAC, SDCWA could supply water to more than 150,000 homes per year (*San Diego Union-Tribune* 2004). In this era, where such powerful interests are at play, the transboundary nature of the Colorado River has been ignored, as have the negative consequences of lining the AAC, not only for Mexico, but also for the habitat of the area.

However, it seems that the current vision of water markets in and of itself would not impede the acknowledgement of transboundary bonds. In their discussion of water, Blatter, Ingram, and Levesque state that negotiations between the political stakeholders—including crossborder ones—involved are possible, although they acknowledge the advantage held by stakeholders who have imposed their vision, and therefore limitations, on the others (Blatter, Ingram, and Levesque 2001). SDCWA consultations with Mexico on the possible construction of a binational aqueduct exemplifies the possibilities of transboundary negotiations. Interestingly, SDCWA is the agency in charge of actually lining the AAC. This leaves open the possibility of a linkage negotiation, at least as far as the modalities of the lining and/or the nature of compensation to Mexico for water losses caused by this lining project are concerned.

On the other hand, other visions of water could feed a new discourse and new actions. Looking into the canal, one can see how the perception of water has been associated with security and has involved the participation of high-ranking officials in the United States and Mexico. The Fox Administration made mention of water as a security tool in connection to both the Río Grande and the Colorado River. With regard to the United States, even in the age of water markets, the U.S. government can surely get involved to modify decisions, especially dealing with the modalities of lining the AAC and/or compensation to Mexico as a result of the implementation of this project. The U.S. government (at the state and federal levels) did not hesitate to delve into the "free market" to achieve the desired IID-SDCWA water transfer agreement, which is considered an important precedent for water-related public policies. In addition, one cannot dismiss the possibility of other water visions coming forth. To that end, it would be worthwhile for non-governmental organizations in both countries to raise their voices against the effects that lining the AAC will undoubtedly have, and demand action be taken by both governments.

In summary, the complex history of the Colorado River in U.S.-Mexican bilateral relations and the role played by a wide variety of governmental and non-governmental agents on both sides, suggest that the AAC is not exclusively a matter of U.S. public policy. The change would require, however, a new discourse and new actions by governmental and non-governmental stakeholders in Mexico and the United States. The purpose would be to rescue the principle of permeability of water flows between California and Baja California, and in any event set a precedent for the many pending debates on transboundary water resources.

References

Álvarez de Williams, A. 1975. *Travelers among the Cucapa* Los Angeles: Dawson's Book Shop.

Associated Press. 1998. "Southern California Water Agencies Approve Transfer Agreement." (10 November).

Blatter, J., H. Ingram, and S. Lorton Levesque. 2001. "Expanding Perspectives on Transboundary Water." In *Reflections on Water: New Approaches to Transboundary Conflicts and Cooperation*, J. Blatter and H. Ingram, eds. Cambridge, Mass.: MIT Press.

Blatter, J., H. Ingram, and P. M. Doughman. 2001. "Emerging Approaches to Comprehend Changing Global Contexts." In *Reflections on Water: New Approaches to Transboundary Conflicts and Cooperation*, J. Blatter and H. Ingram, eds. Cambridge, Mass.: MIT Press.

Boyarsksy, B. 1997. "Swimming with the Big Fish in the Water War." *Los Angeles Times* (8 September): B-1.

Brazil, E. 2001. "Environment Colorado River 'Surplus' Challenge." *San Francisco Chronicle* (9 December): 1.

Bureau of Citizenship and Immigration Services. 2001. "U.S. and Mexico Pledge to Expand Safety Efforts Along California Border." (22 June). http://www.immigration.gov/graphics. publicaffairs/newsrels/binatmtngrel.htm.

California Senate. 1998. "SB 1765." http://info.sen.ca.gov/pub/97-98/bill/sen/sb_1751-1800/sb_1765_bill_19980925_chaptered.html.

California Senate. 2003a. "SB 277." http://info.sen.ca.gov/pub/03-04/bill/sen/sb_0251-0300/sb_277_bill_20030929_chaptered.html.

California Senate. 2003b. "SB 317." http://info.sen.ca.gov/pub/03-04/bill/sen/sb_0301-0350/sb_317_bill_20030929_chaptered.html.

California Senate. 2003c. "SB 654." http://info.sen.ca.gov/pub/03-04/bill/sen/sb_0651-0700/sb_654_bill_20030929_chaptered.html.

Conaughton, G. 2001. "San Diegans Oppose Plan to Pipe Colorado River Water through Mexico." *North County Times* (10 February).

Cortez Lara, A., and M. R. García Acevedo. 2000. "The Lining of the All-American Canal: The Forgotten Voices." *Natural Resources Journal* 40 (2).

Dibble, S. 2004. "Mexican Opposition to Canal Lining Grows." *San Diego Union-Tribune* (18 December): 1.

DiMento, J. 2001. "Black Sea Environmental Management: Prospects for New Paradigms in Transitional Contexts." In *Reflections on Water: New Approaches to Transboundary Conflicts and Cooperation*, J. Blatter and H. Ingram, eds. Cambridge, Mass.: MIT Press.

Doughman, P. M. 2001. "Discourses and Water in the U.S.-Mexico Border Region." In *Reflections on Water: New Approaches to Transboundary Conflicts and Cooperation*, J. Blatter and H. Ingram, eds. Cambridge, Mass.: MIT Press.

Fradkin, P. L. 1981. *A River No More*. New York: Alfred A. Knopf.

García Acevedo, M. R. 2001. "The Confluence of Water, Patterns of Settlement, and Constructions of the Border in the Imperial and the Mexicali Valleys (1990–1999)." In *Reflections on Water: New Approaches to Transboundary Conflicts and Cooperation*, J. Blatter and H. Ingram, eds. Cambridge, Mass.: MIT Press.

Gardner, M. 2001. "State Energy Spending Gobbles Critical Water Program Funding." *Copley News Service* (2 February).

Gottlieb, R. 1988. A Life of its Own. *The Politics and Power of Water*. San Diego: Harcourt Brace/Javanovich Publishers.

Gottlieb, R., and M. Fitzsimmons. 1991. *Thirst for Growth. Water Agencies as Hidden Government in California*. Tucson: University of Arizona Press.

Hayes, D. 1991. "The All-American Canal Lining Project: A Catalyst of Rational and Comprehensive Groundwater Management on the U.S.-Mexico Border." *Natural Resources Journal* 31.

Henderson, T. 1968. *Imperial Valley*. San Diego: Neyenesch Printers Inc.

Hume, B. 2000. "Water in the U.S.-Mexico Border." *Natural Resources Journal* 40 (2).

Ingram, H. 2000. "Transboundary Groundwater on the U.S.-Mexican Border. Is the Glass Half Full, Half Empty or Even on the Table?" *Natural Resources Journal* 40 (2).

International Boundary and Water Commission. 1973. *Minute 242: Permanent and Definitive Solution to the International Problem of the Salinity of the Colorado River*. http://www.ibwc.state.gov/Files/Minutes/Min242.pdf.

Jiménez, J. L. 2003. "County Water Authority Approves Imperial Deal." *Los Angeles Times* (26 September): B2-B3.

Jones, L. C., P. Duncan, and S. P. Mumme. 1997. "Assessing Transboundary Environmental Impacts on the U.S.-Mexican and U.S.-Canadian Borders." *Journal of Borderlands Studies* XII (1): 74–96.

Kerig, D. P. 1988. "Yankee Enclave: The Colorado River Land Company and the Mexican Agrarian Reform in Baja California." Ph.D. dissertation, University of California, Irvine, Irvine, California.

LaRue, S. 2001a. "Utility Companies to Study Aqueduct Plan; Cross Border Pipeline Proposed." *San Diego Union-Tribune* (26 January).

LaRue, S. 2001b. "Water Access to be Restored to 5 N. County Indian Bands." *San Diego Union-Tribune* (3 February).

Leavenworth, S. 2003. "Ink Drying on Western Water Pact At Long Last; San Diego's Thirst Slaked; Imperial Fears Abated; Bureaucrats Taking Bows." *Sacramento Bee* (11 October): A1.

Los Angeles Daily News. 1996. "Conservation Pays. MDW's Plan to Save Water Deserves Support." (4 January): 1.

McClurg, S. 1996. "Colorado River Controversies." *Western Water* March–April.

McKinnon, S. 2000. "Babbitt Settles Feud with California; Western States Protect their Share of the Colorado River." *The Arizona Republic* (17 January).

Metz, L. C. 1989. *Border: The U.S.-Mexico Line*. El Paso: Mangan Books.

Mumme, S. P. 1996. "Groundwater Management on the Mexico-United States Border." Unpublished manuscript.

Mumme, S. P. 2000. "Minute 242 and Beyond: Challenges and Opportunities for

Managing Transboundary Groundwater on the Mexico-U.S. Border." *Natural Resources Journal* 40 (2).

Nolde, H. 2001. "El Canal Todo Americano ya no regará los campos mexicanos." *La Opinión* (25 February).

Office of the Governor of California. 2003. "Governor Davis Signs Unprecedented Water Legislation that Implements Historical Accord." (29 September).

Office of the Presidency of the Republic. 2001. "Entrevista al Presidente Vicente Fox en el Rancho San Cristobal." (18 February). http://www.presidencia.gob.mx.

Pearson, R. P. 1996. "California Top Water Bank Could Shortchange Arizona." *The Arizona Republic* (28 January).

Perry, R., J. Blatter, and H. Ingram. 2001. "Emerging Approaches to Comprehend Changing Global Contexts." In *Reflections on Water: New Approaches to Transboundary Conflicts and Cooperation*, J. Blatter and H. Ingram, eds. Cambridge, Mass.: MIT Press.

Perry, T. 1997. "Southland Water Future May Hinge on Water Dispute." *Los Angeles Times* (3 August): A-1.

Perry, T. 2000. "California and the West; Calming Waters." *Los Angeles Times* (10 December).

Perry, T. 2003a. "Legislation Seeks End of Water War." *Los Angeles Times* (30 September): A1.

Perry, T. 2003b. "Imperial Water Deal Completed." *Los Angeles Times* (3 October): B1.

Postel, S., P. Gleick, and J. Morrison. 1996. *The Sustainable Use of Water in the Lower Colorado River Basin.* Oakland: Pacific Institute for Studies in Development, Environment and Security.

Raley, B. 2001. Testimony before U.S House of Representatives Subcommittee on Water and Power. http://resorcescommittee.house.gov/107cong/water/2001dec10/raley.htm.

Restrepo, I. 2000. "El Canal Todo Americano." *La Jornada* (8 May).

Ritter, K. 2001. "Federal Official Warns California Must Curb Colorado Water Use." *Associated Press* (13 December): 1.

Rohrlich, T. 1998. "Thirst to Overhaul Powerful Water Agency Grows Stronger." *Los Angeles Times* (19 July): B-1.

San Diego County Water Authority. 1997. "Fact Sheet: Water Conservation and Transfer Agreement San Diego County-Imperial Valley." (11 December). http://www.sdcwa.org/text/dfaq.htm.

San Diego County Water Authority. 1998a. "Landmark Water Conservation and Transfer Agreement Ratification." (28 April). http://www.sdcwa.org/text/pressrel/iidvote2.htm.

San Diego County Water Authority. 1998b. "Bill to Fund California 4.4 Plan Approved by Legislature." (1 September). http://www.sdcwa.org/text/pressrel/sb1765.htm.

San Diego County Water Authority. 1999. "Mexico, U.S. Sign Agreement to Launch Aqueduct Feasibility Study." (14 October).

San Diego County Water Authority. 2001. "Secretary of the Interior Babbitt to Sign Historic Colorado River Agreement." 12 January.

San Diego County Water Authority. 2003. "Four Agencies Sign Historic Colorado River Deal Quantification Settlement Agreement Secures New, Reliable Water Supply for San Diego County." (22 October). http://www.sdcwa.org/news/101003AgenciesSignQSA.phtml.

San Diego Union-Tribune. 2004. "Line the Canal: Water Transfer Deal Must Go Forward." (26 December): 1.

Sánchez Ramírez, O. 1990. *Crónica agrícola del Valle de Mexicali*. Mexicali: Universidad Autónoma de Baja California.

Sierra Club. 2001. "Colorado River Delta Restoration." (10 April). http://www.sierraclub.org/rcc/southwest/coreport/recommendations.asp#.

Sierra Club Regional Conservation Committees. 2002. "Colorado River Report." (3 March). http://www.sierraclub.org/rcc/southwest/coreport/recommendations.asp.

Snedeker, L. 2000a. "Interior Secretary Announces Plan for the Colorado River Water." *Associated Press* (14 December).

Snedeker, L. 2000b. "Agreement Doesn't Give Water Away to Mexico." *Associated Press* (15 December).

Stapleton, M. 2001. Testimony before U.S. House of Representatives Subcommittee on Water and Power. http://resorcescommittee.house.gov/107cong/water/2001dec10/stapleton.htm.

Secretaría de Relaciones Exteriores. 1975. *La Salinidad del Río Colorado*. México, D.F.: Colección del Archivo Histórico Diplomático Mexicano.

The Washington Post. 2000. "U.S., Mexico Probe Reported Border Breach." (16 July).

U.S. House of Representatives. 2001. "HR 2764." http://thomas.loc.gov/cgi-bin/bdquery/z?d107:H.R.2764:.

U.S. House of Representatives, Committee on Interior and Insular Affairs. 1988. *Lining the All-American Canal*. Washington, D.C.: U.S. Government Printing Office.

U.S. Senate. 1988. "S. 795" (became Public Law 100-675). http://thomas.loc.gov/cgi-bin/bdquery/D?d100:7:./temp/~bdVYYj::.

Utton, A. E. 1991. "The Transfer of Water from an International Border Region: A Tale of Six Cities and the All-American Canal." *Natural Resources Journal* 16: 477–490.

Walters, D. 2000. "Babbitt Leaves the Glass Half Full." *San Diego Union-Tribune* (21 December): B-14.

Ward, E. 1999. "Two Rivers, Two Nations, One History: The Transformation of the Colorado River Delta since 1940." *Frontera Norte* 11: 113–140.

VII

Beyond the All-American Canal: Future Scenarios of Water Availability in the Mexicali Valley

José Luis Castro Ruíz

Introduction

In 1988, Congress passed House Bill 100-675 authorizing the U.S. Department of the Interior to order the lining of the All-American Canal (AAC) in California's Imperial Valley. It would recover approximately 100,000 acre-feet (AF) of water lost annually through seepage along the 60 kilometers (km) (40 miles) of the canal that run parallel to the U.S.-Mexican border. In December of the same year, the Imperial Valley Water District published its *Water Conservation and Implementation Plan,* which established the lining of the main canals that transport water from the Colorado River as an essential measure to achieve the utility's expected water savings.

The unilateral nature of these events not only upset and concerned Mexican officials but launched a debate about how lining the AAC would affect Mexico. It was a debate that, until recently, carried on mainly in academic and non-governmental organization circles and focused on the project's implications for the agricultural users on the Mexican side of the canal's area of influence. This area has been estimated at nearly 19,182 hectares (47,400 acres) (Cortez Lara and García Acevedo 2000) and is located in the northwest portion of the Mexicali Valley.

The issue of AAC lining has been raised again, given the U.S. government's decision to proceed with the project despite the Mexican position about the then-existing water debt from the Rio Grande. The arrival of this long-foreseen scenario has renewed interest in the project's impacts.

This chapter provides context for the concerns that those managing and planning water resources in Baja California have about urban water demand. The first of three sections of this chapter provide background on the state's urban development and in particular development of the Mexicali Valley, as well as how this development has progressed given low water availability. In the second, the current conditions of the water supply in the region are addressed. Following this, some hypotheses on urban demand scenarios are posed and discussed, both for the valley itself as well as the state in general. Any scenarios that deal with lining the AAC must consider the restrictive impacts of the available water supply.

Baja California: Urban Development and Water Dependency

The concentration of population near the water supply has been characteristic of the development of the border region between Mexico and the United States. Compared to other areas along the border, the urbanization of the Baja California border was a late but dynamic phenomenon defined in large measure by its historic relationship with the state of California. With the exception of the settlements located in the Mexicali Valley, the problem of water availability for Baja California communities has been endemic.

Urban growth in this region began in the early 20th century. Then, Tijuana and Tecate were small settlements of 242 and 127 inhabitants, respectively,[1] and provided minimal services for the transient population. Events such as the surge of development in the Mexicali and Imperial Valleys, and later Prohibition in the United States, generated the initial conditions for urban expansion that the border would exhibit in the decades that followed. As Table 1 shows, by 1921 Mexicali had consolidated itself as a service center for agricultural activities in the northeastern part of the state. In Tijuana the balance of the Prohibition years were characterized by annual population growth rates of 26.3% between 1921 and 1930, with migrants accounting for nearly 95% of that growth rate. These conditions began to generate a water availability problem for the city that would continue for the next 50 years.[2]

The *Bracero* guest worker program during World War II and the dynamic reputations these border cities had acquired further triggered development in the following decades. The 1940s marked the beginning of a stage of sustained growth that would last through the second half of the 20th century and that the cities of Ensenada and Tecate would join in time.[3] Upon the end of the *Bracero* program in the 1960s, the Mexican government promoted the Programa Industrial Fronterizo (PIF, in English Border Industrial Program) which provided an alternative for the border cities' labor market. Although the consolidation of this program in the Baja California border region was not immediate, eventually its impact transformed the economic base and the urban structure of cities such as Tijuana.

Currently, with the exception of Mexicali, all the cities in the state maintain annual growth rates in excess of 3.0%. The influence of the maquiladora sector, in both the municipalities where they first located and others where the sector diversifies the economy, continues to be a basic determining factor. The municipality of Tijuana has 48.7% of the state's population, and more than 95% of residents live in the city of Tijuana itself. As a whole, by 2000 Baja California's 39 border municipalities held 36.9% of the population.

Mexicali and Its Valley

The urbanization of the Mexicali Valley has been intimately linked to the rise and evolution of the agricultural sector in the region. This entire development was sustained by the Colorado River, given its strategic proximity and seasonal reliability (Figure 1). While the city gained popularity as a central border crossing point from its founding in the summer of 1901, it did not boom until 1907 when flooding of the arable land, caused by failures in the canal systems, was resolved. This, in turn, allowed for production in the Imperial

Table 1. Population Growth of the Principal Cities in Baja California, 1900–2000

	1900	1910	1921	1930	1940	1950	1960	1970	1980	1990	2000
Tijuana	242	733	1,028	8,384	16,486	59,952	152,473	277,306	461,257	702,228	1,148,681
Census Rate (%)	-	11.7	3.1	26.3	7.0	13.8	9.8	6.2	5.2	4.3	5.0
Mexicali	-	462	6,782	14,842	18,775	65,749	179,539	263,498	341,559	439,756	549,873
Census Rate (%)	-	-	27.7	9.1	2.4	13.4	10.6	3.9	2.6	2.6	2.3
Ensenada	1,726	2,170	2,178	3,042	4,616	18,150	42,561	79,146	120,483	259,379	370,730
Census Rate (%)	-	2.3	0.3	3.8	4.3	14.7	8.9	6.4	4.3	8.0	3.6
Tecate	127	-	493	566	-	3,679	7,074	14,738	30,540	51,946	77,795
Census Rate (%)	-	-	6.7[1]	1.5	-	9.8[2]	6.8	7.6	7.6	5.5	4.1
Rosarito	-	-	-	-	-	-	-	-	-	23,067	49,178
Census Rate (%)	-	-	-	-	-	-	-	-	-	-	7.9

Valley and the eventual consolidation of an important regional labor market on both sides of the border. This led to settlement in the city by small farmers, business people, and providers of basic services, who in turn attracted new immigrants employed in irrigation and communications projects. The concession policies of the Porfirio Díaz Administration that characterize this era promoted U.S. investment in the Mexicali Valley, which led to the establishment of the Colorado River Land Company. The company's presence determined the control of the economic activities in the valley through its financial and land-leasing operations, as well as its experimental planting of cotton (Anguiano 1995). By 1910, the population of Mexicali and its valley totaled 1,600, and approximately one-third of these people lived in the city and worked in the service sector. Prior to the 1930s, the city's economy consolidated around the services generated by cotton production in the valley and the emerging market opened up by Prohibition in the United States. During those years, immigration into the valley and the city was intense and resulted in Mexicali's having the greatest concentration of the state's population (62%). The city's population grew to more than 14,000 between 1910 and 1930 (see Table 1).

The 1930s marked a series of changes for the valley's economy. First, the worldwide economic crisis of the times caused a drop in cotton production, which led to an 80% drop in single-crop production. Then, changes in national policy regarding the restructuring of territorial property promoted the expropriation and redistribution of large areas of land held by Colorado River Land. Finally, and as complementary measures by the Lázaro Cárdenas Administration to support the *ejidatarios* (communal land owners) and colonizers, the Distrito de Riego del Río Colorado (in English, Colorado River Irrigation District) was created in 1938, rising from the displacement of Colorado River Land's control of the water in the valley. This period, essential for the coming years and characterized by a drop in economic activity, nevertheless did not affect the municipality's general growth. Population increased through new immigration, attracted by the federal government's agrarian policies and the creation of *ejidos* as a basis for agrarian distribution.

Figure 1. Principle Urban Settlements in Mexicali Valley

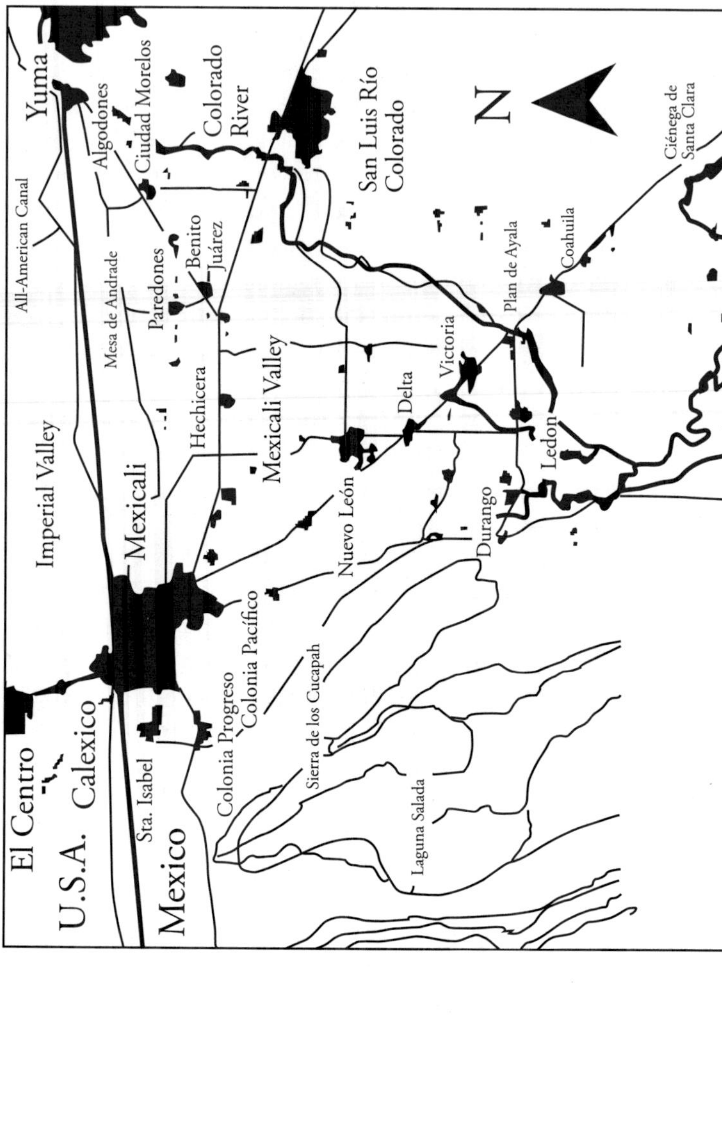

During the 1940s and 1950s the valley's economy consolidated around cotton. After a period of depressed fiber prices all over the world during the war years, production finally rebounded. Other subsequent events, such as the Korean War and the devaluation of the Mexican peso, helped increase the supply of cotton even more. In the Mexicali Valley, all these events occurred at the same time as an increase in irrigation and communications infrastructure investment including: the construction of the Morelos Dam to divert and store the water from the Colorado River under the 1944 Water Treaty,[4] the expansion of the network of canals in the valley, the construction of the Sonora-Baja California railroad that finally united the peninsula with the rest of the country, the construction of highways to the Sea of Cortez and the town of San Luis Río Colorado, and the construction of a network of rural roads uniting the valley with the municipal seat. During this period, the populations of both the city and the municipality grew (Table 2) as a result of inter-municipal migration as well as the arrival of immigrants from other parts of the country, attracted by the *Bracero* program.

The 1960s marked the beginning of the last stage in the valley's urban growth. It was during these years that PIF began, and while it is true that it started slowly in Mexicali and other cities, its impact on the economic base has been significant in the last 20 years[5] and eventually changed the relative participation of the municipal population.[6] Some settlements in the valley became population magnets, in some cases showing growth rates higher than the municipality's (Table 3). Currently, more than 80% of the municipality's urban population is concentrated in the Mexicali Valley alone.

Table 2. Mexicali's Population and Growth Rates, 1930–2000

Year	Municipality	Annual rate (%)	City	Annual rate (%)	Percent Population in City
1930	29,985	n/a	14,842	n/a	49.5
1940	44,399	4.0	18,775	2.4	42.3
1950	124,362	10.8	65,749	13.4	52.9
1960	281,333	8.5	179,539	10.6	63.8
1970	396,324	3.5	263,498	3.9	66.5
1980	510,664	2.6	341,559	2.6	66.9
1990	601,938	1.7	439,756	2.6	73.1
2000	764,602	2.4	549,873	2.3	71.9

The Colorado River and Urban Uses in Baja California

Estimates show that Baja California has a total annual supply of 2.52 million acre-feet (MAF) of water, of which 1.5 MAF (60%) comes from the only reliable surface flow in the state—the water allocation from the Colorado River that the United States delivers to Mexico.[7] The remaining 1 MAF is extracted from the region's aquifers (Comisión Estatal del Agua de Baja California 2003). The Colorado River water delivered to the Mexicali Valley totals 71% of the 2.1 MAF available annually for its urban and rural uses, including agriculture.[8] This volume alone is 85.5% of the total annual supply of Baja California's statewide needs (Román and Bernal 1995), which is why the true availability of this resource is unknown—it could be transferred to other regions of the state where it might be needed. Currently, the Comisión de Servicios de Agua del Estado (COSAE, in English State Water Services Commission)[9] receives from Comisión Nacional del Agua (CNA, in English National Water Commission) an annual volume of 162,000 AF from the Colorado River quota and approximately 51,000 AF

Table 3. Urban Locations in the Mexicali Valley

Locale	Population 1990	Population 2000	Annual Inter-census Rate
Mexicali	439,756	549,873	2.3
Santa Isabel	5,624	18,041	12.4
Guadalupe Victoria	10,816	15,561	3.7
Puebla	5,420	7,421	3.5
Ciudad Morelos (Cuervos)	6,031	7,234	1.8
Alberto Oviedo Mota	6,279	6,878	0.9
Estación Coahuila	5,349	6,479	1.9
Hermosillo	4,974	5,458	0.9
Estación Delta	5,040	4,860	-0.4
Benito Juárez (Tecolotes)	4,242	4,486	0.6
Progreso	2,457	4,462	5.8
Vicente Guerrero (Algodones)	3,489	4,157	1.8
Paredones	3,510	3,634	0.3
Nuevo León	3,073	3,255	0.6
Michoacán de Ocampo	2,659	3,237	2.0
Mexicali Valley (*)	68,873	95,163	3.3
Municipio of Mexicali	601,938	764,602	2.4
Baja California	1,660,855	2,487,367	4.1

Note: Does not include the city of Mexicali.

from aquifers to satisfy the urban demand in the state. Approximately half is transferred from Irrigation District 014 to the cities of Tecate and Tijuana via the Colorado River-Tijuana Aqueduct (in Spanish, ARCT). The total volume extracted from wells includes the various aquifers in the state.

Of the 81,000 AF of Colorado River water remaining from COSAE's allocation for urban demands, nearly 82% is transferred to the Comisión Estatal de Servicios Públicos de Mexicali (CESPM) to cover the municipality's needs.[10] CESPM must complement current demand with 13,300 AF from the purchase of water rights from farmers in the valley and 3,500 AF from other sources.[11] With this supply of water, CESPM currently services nearly 98% of the city's population and 33% of the remainder of the municipality's population (CESPM 2000). By the end of 2000, CESPM registered a total of 20,215 installed connections with which it serves approximately 100 settlements in the remainder of the municipality,[12] including the largest settlement in the valley (see Table 3).

The volume of water transferred via ARCT to Tecate and Tijuana totals 72% and 95%, respectively, of each area's annual need. The Comisión Estatal de Servicios Públicos de Tecate complements its urban water needs with water from the aquifers of the city and the Arroyo San José area. The city of Tijuana has other supply sources, but their relative contribution is quite low. These include the Abelardo L. Rodríguez Dam (1%) and wells located in the riverbed of the Tijuana River and in the la Misión and Rosarito areas (4%).

Future Water Demand in the Valley and State: Some Hypotheses

The city and valley of Mexicali have historically been perceived as a region removed from the water availability problems of the state of Baja California by virtue of their proximity to the Colorado River. Undoubtedly this attitude has permeated the water consumption and administration patterns of the various consumer sectors. Notably, CESPM emphasizes in medium- and long-term planning alternatives that the water allocations to the valley's sectors will not change in the future.[13] In contrast with these expectations, both the water use trends previously outlined as well as the high proportion of water resources concentrated in this area suggest the need for more focus on the demand scenarios that may arise in the future. To this end, the following section poses two feasible scenarios of future water demand, both for the valley itself and the state in general, and includes the effects that lining the AAC could have on water supply planning in each case.

Mexicali and Its Surroundings

The conditions that may arise in the valley in the medium- and long-term warrant the most immediate analysis. One obvious hypothesis is that, to the extent urban growth in the city and valley maintain at least a similar pace to what has been experienced to date, important urban settlements will continue to consolidate and attract migration to the area.

Official estimates for Mexicali foresee a population of 1.1 million inhabitants by 2020, an increase of approximately 100% of the population reported in the 2000 census (CNA 2000). With a conservative assumption that the current allocation of 117 gallons of water per day per inhabitant reported by the city is maintained, the population growth will generate an approximate annual demand of 143,500 AF. This requires, independent of the needs of other cities in the state, an annual increase in the urban water supply of 77,000 AF. For the rest of the municipality, the same estimation parameters show an increase in the population by 2020 of more than 175,000 inhabitants. Assuming the same proportion of urban residents currently concentrated in the valley will be maintained, the result would be approximately 149,000 new residents with an approximate additional annual consumption of nearly 19,000 AF.

While the demographic dynamic is the variable here, the effects of the growth and diversification of the business sector cannot be ignored. The location of industrial plants and the expansion of commerce and services also increase water demand. CESPM reported for 2000 a growth of water consumption in the non-domestic sector of slightly more than 800 AF annually (CESPM 2000). Considering an incremental increase of this nature in the non-domestic sector, the result would be an additional urban demand of 16,200 AF in 2020.

The foregoing is a scenario of urban water demand raised here only for discussion purposes. It assumes that the current trends of population growth, urban economic activities, and the consumption patterns among the different sectors will be maintained in the long run. Nevertheless, this framework brings up such questions as:

- What are the implications of this panorama on the valley's water supply?

- What effects will this increase in urban demand have on the valley's agricultural activities?

Returning to the estimated numbers, the assumed scenario will result in an additional annual demand of 112,600 AF of water, a volume double CESPM's reported production in 2000. It is reasonable to suppose that the responsibility for supplying this increase will fall on the utility, considering the goal of 100% service coverage in a relatively shorter time frame than the 2020 horizon. Assuming for these purposes that the same proportional scheme of supply sources will be maintained, CESPM will need an additional re-allocation of nearly 800,000 AF of Colorado River water, which is equal to the quota CNA currently receives to cover the urban demand of the entire state.

The previous amount would still need to be increased by nearly 17,800 AF and acquired through the purchase of irrigation rights from farmers in the valley. The trend this process follows will depend a great deal upon how the city expands spatially, as well as upon the willingness of the irrigation users to sell their rights.[14] In any event, a transfer of these characteristics will accelerate the change from mostly agricultural to urban water consumption in the valley.

Lining the AAC

The lining of the AAC cannot be ignored when considering future water demand in the Mexicali Valley. Studies of the seepage from the AAC into the valley's aquifer show an annual recharge of nearly 64,800 AF (Secretaría de Recursos Hidráulicos [SRH] 1972). Technically, this volume is part of the aquifer's annual 632,300 acre-foot contribution to the total available water production in the valley. Studies on the effects of lining the AAC have emphasized the changes in the quantity and quality of agricultural irrigation water in the valley as an effect of the high number of users supplied through wells.[15] In the case of the water volumes related to the urban uses in the valley, its management falls primarily on CESPM, which basically establishes its supply with contributions from the Colorado River.

Taking into consideration the current supplies upon which CESPM relies, it can be assumed that lining the AAC will not have an immediate effect on the water supply available to the city. The relevant scenario for this discussion will take place on a much larger scale, though, when the eventual increase of demand forces the utility to seek other options, including using aquifer water. This change will not only establish new pressures on resources that have traditionally been assigned to agriculture, but a reduction in the recharge volumes as a result of lining the AAC in the area immediately north of the Mesa Arenosa (Sandy Mesa) could accelerate conflict between urban and agricultural uses in the valley. This scenario will require significant attention by the utility to either negotiate access to greater volumes from the Colorado River or seek support to give it priority over non-urban uses in accessing aquifer water. These conditions could occur with the allocation of the available water CESPM has in Mesa de Andrade, which it does not currently access because it needs to build an aqueduct but which nevertheless forms part of the utility's future expansion plans.

Two other elements should likewise be considered. The first deals with the quota CNA delivers to the municipality of San Luis Río Colorado from the Mesa Arenosa area, which at present amounts to 16,200 AF annually (CNA 2000). Assuming this municipality maintains the growth rate of the past two decades into 2020, water demand will increase by approximately 4,900 AF. In the event it cannot satisfy delivery of this volume, CNA will have to seek other alternatives, among them Colorado River water. The second element has to do with the levels of consumption by other settlements in the valley not served by CESPM and whose supply options will continue to depend greatly upon the aquifer. For these groups the effect will not only be a reduction in available water volumes, but also the deterioration of its quality.

Water Demand in the State

The scenarios discussed thus far have been limited to those with internal impacts on the Mexicali Valley as a result of population growth and the evolving dynamic of economic activity. There is, however, a longer-term condition whose feasibility is increased if the state's urban population grows, particularly in the border region—the expectations Colorado River water will generate in urban centers located beyond the valley. To this end, consider on the one hand that the quota from this source is fixed, and on the other hand that the official population projections for this region will increase by 100% by 2020 (CNA 2000). A scenario of open competition for Colorado River water, which could lead to conflict between the parties, cannot be ignored.

Independent of any water saving and wastewater reclamation measures considered by state water management programs in the medium- and long-term, the supply proposals are strongly based on the possibility of expanding, little by little, the volume of Colorado River water transferred from the Mexicali Valley. Currently, the design capacity of ARCT is 1,190 gallons per second (g/s) (Comisión Estatal de Servicios Públicos de Tijuana 2003), which results in an increase in the current transfer of water of some 30,800 AF annually to fulfill the demand of the next five years. By that time, multiple proposals will depend on Colorado River water not only to ensure the supply for cities currently using it but also to increase coverage to include Ensenada. For example, the Plan Estatal Hidráulico 1994–2015 (in English, State Hydraulic Plan) developed by COSAE, establishes as part of its proposals a series of supply scenarios during this period that call for increased use of Colorado River water through the construction of ARCT II in 2006. Other proposals include increasing the aqueduct's capacity from 1,050 g/s to 1,400 g/s with certain modifications, such as two long-term options that would ensure a delivery of up to 1,850 g/s for Baja California's northwest region.[16]

Figure 2 shows the historic relationship between the Colorado River water supply and the state's urban population served from the signing of the 1944 Water Treaty when the allocation of 1.5 MAF annually for Mexico was established. Figure 2 also incorporates the urban population forecasts previously discussed, as well as the Colorado River water supply coverage objectives for the state's urban centers in the next 20 years. The information shown there provides an idea of the effects relevant policies to increase supply coverage had on Colorado River water availability. One important change in this sense occurred in the 1980s when ARCT began operating, adding the cities of Tecate and Tijuana to the served population. Another important turning point is expected to take place in this decade, when Ensenada will have access to Colorado River water via the expansion of the Morelos aqueduct to the area known as El Hongo.

Some Conclusions

The ultimate objective of this chapter has been to transcend in some measure the lining of the AAC itself and instead draw attention to a fundamental state problem whose long-term solution is uncertain: Urban water demand. This discussion of hypothetical scenarios in no way pretends to be definitive in relation to the future of water resources in the state. The elements addressed in this chapter outline the complexity that will continue to surround decision-making on this issue. For the time being, this reflection has identified some critical points that should garner attention in light of the future of water availability in the state. These are described below.

1. The lining of the AAC will affect the availability of urban water in the Mexicali Valley because future alternatives slowly incorporate aquifer water.

Figure 2. Water Supply vs. Population

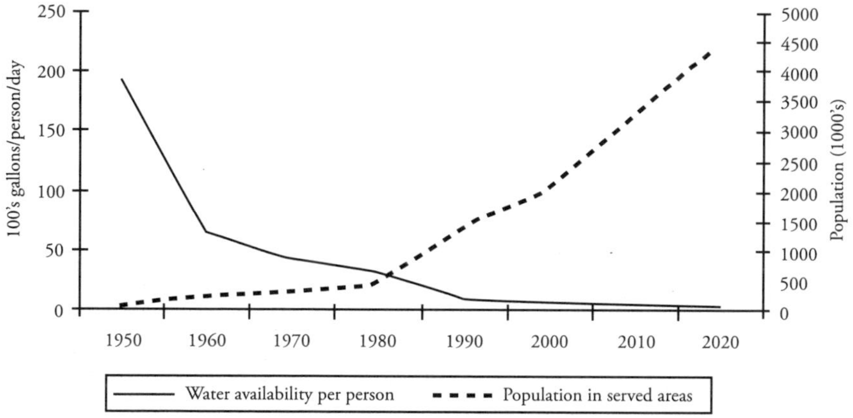

The most important volume of water for urban uses in the Mexicali Valley is the volume managed by CESPM. The present origin of this volume assumes that the AAC project will not have significant short-term effects. However, this scheme will need modification when the utility uses any allocation of underground water, because this would imply an additional load that will have to be addressed with a supply that, in reality, would be lower because of the AAC lining. The project contemplated by CESPM to access its original water allocations from CNA in the Mesa de Andrade offers an interesting opportunity to test and prove (or disprove) the described conditions.

2. The valley's use structure may undergo profound changes in the future.

The utility's present practice of purchasing water rights may acquire the character of an official resource assignment mechanism through institutional changes. This would eventually accelerate the process of change in water use from non-urban to urban activities. Moreover, the formal existence of a water marketplace between uses in the state could facilitate the transfer of Colorado River water from the valley's agricultural use to other regions of the state that may have an urgent need for it.

3. Colorado River water will continue to be the basis for future supply expectations of the region's governments.

Though supply-and-demand scenarios posed here have a marked dependence on Colorado River water. At this time, discourse at different levels of government in the state already includes the consideration of alternative measures to deal with future requirements, such as paying particular attention to maintenance programs to reduce losses in the potable water network, the reclamation and reuse of treated wastewater, or the promotion of programs to raise public awareness of water conservation. In spite of the foregoing, these elements will still require a long process to come to fruition—a process that includes considerable investment yet can nonetheless compete with the investments required to increase Colorado River water volume.

4. The problem of future water availability in the state must be addressed, and officials must act in an integral, all-inclusive manner.

One obvious conclusion is that the state needs a new process for prioritizing the available water supply. The planning and management of water resources in Baja California has had a history of delayed actions on the one hand, and a lack of concern about its availability on the other. Currently, the lack of concrete planning actions to confront the long-term demand conditions by different levels of government seems to be a partial result of this historic inertia. Certainly the lining of the AAC is one more aggravating element in the uncertain panorama the state faces to ensure the supply of water for generations to come, but this event could also be a catalyst if it increases awareness within the governments and parties involved in the management and planning of the resource. It could lead them together to act collectively while bearing in mind that the water supply will at best remain fixed for the coming decades and that the strategic importance of the Mexicali Valley must be understood and given preference over local planning proposals.

Endnotes

[1] Despite the fact that Tijuana owed its surge to the establishment of the new border between Mexico and the United States in 1848, its distance and isolation from the rest of the country did not allow it to grow beyond a modest settlement for more than 50 years (Arreola and Curtis 1993).

[2] As opposed to Mexicali, whose geographic proximity to the Colorado River historically allowed it to ignore water supply problems, Tijuana's dependency on seasonal flows like the Tijuana River and available underground sources was a permanent catalyst to search for options to meet the growing urban demand. The completion of the Abelardo L. Rodríguez Dam in the mid-1930s generated limitless mid- and long-term expectations for the city, which plummeted with a prolonged drought in the region the following decade (Tapia 1989).

3 The agricultural sector was an important promoter of growth for the city of Ensenada after its beginnings as a Customs port of entry at the end of the 19th century. Since the 1940s the city has shown sustained growth, which continues to this day. Throughout this process, authorities have moved from total dependence on subterranean sources close to the urban area to a combination of seasonal surface flows that are captured in the Emilio López Zamora Dam, as well as on water coming from more distant aquifers, such as the Guadalupe Valley and from Maneadero, whose volumes are conveyed via pipelines.

4 The Treaty on the Utilization of the Bravo, Colorado and Tijuana Rivers, was signed by the governments of Mexico and the United States on February 3, 1944.

5 The number of direct jobs in the *maquiladora* sector of Mexicali increased from 10,876 in 1975 to nearly 25,000 in 1995 (Lorey 1999).

6 There have been other federal mechanisms whose effects have undoubtedly contributed to the city's urban development in the direction implied here, especially in the commercial sector. Among these are the Programa Nacional Fronterizo (PRONAF, in English National Border Program) in 1961 and the "loss-leader articles" program put in place during the Luis Echeverría Administration.

7 This quota was established as part of the 1944 Water Treaty.

8 The Colorado River water is likewise a source for recharging the Mexicali Valley aquifer, whose use provides an additional 570,000 AF to the available supply for the valley's urban and agricultural uses. This source is also the main supply for the neighboring municipality of San Luis Río Colorado, Sonora (CEABC 2003).

9 The Comisión de Servicios de Agua del Estado is a decentralized agency of the state government whose functions include moving water en mass through the aqueduct system to supply the communities and negotiating and coordinating between the diverse agencies responsible for supplying water for urban and rural uses (Castro and Sánchez 2001).

10 CESPM is one of the 760 clients of irrigation module 016 in the Mexicali Valley. As such it receives a direct allocation of Colorado River water from CNA.

[11] This is according to Comisión Estatal de Servicios Públicos de Mexicali (CESPM) (2000).
[12] This information was gleaned in an interview with engineer Luis Manuel Venegas of CESPM on March 29, 2001.
[13] These include continuing to purchase water rights from farmers, or the possibility of reclaiming water that is part of the original CNA allocation to the area known as Mesa de Andrade, east of the city.
[14] Thus, it is quite feasible that within this timeframe both federal and state legislative frameworks will consider the necessary modifications to make water markets possible between different uses in the country.
[15] In what is considered the AAC's area of influence on the Mexican side, 60% of agricultural consumers depend exclusively on aquifer water for their needs. Another 27% depend only on Colorado River water, while the remaining 13% use a combination of both (Cortez Lara and García Acevedo 2000).
[16] For this horizon two options stand out: building a second aqueduct with strictly local funding and building a binational aqueduct that would transport water from the Imperial Valley through Mexican territory to San Diego, California. This was promoted by private industry and could have transfered up to 3,170 g/s for the City of San Diego, as well as complied with delivery goals for Baja California. However, this proposal has since died on the U.S. side.

References

Anguiano, M. E. 1995. *Agricultura y migración en el valle de Mexicali*. Tijuana, B.C.: El Colegio de la Frontera Norte.

Arreola, D. D., and J. R. Curtis. 1993. *The Mexican Border Cities: Landscape Anatomy and Place Personality*. Tucson, Ariz.: The University of Arizona Press.

Castro Ruiz, J. L., and V. Sánchez Munguía. 2001. "Cambio institucional y gestión urbana del agua: El caso de Baja California." Pages 239–262 in *Planeación y gestión urbana y metropolitana en México: Una revisión a la luz de la globalización*, R. García Ortega, ed. Toluca, Edo. de México: El Colegio Mexiquense.

Comisión Estatal de Servicios Públicos de Mexicali. 2000. *Indicadores de gestión 2000, bases de proyección*. Mexicali, B.C.: CESPM.

Comisión Estatal de Servicios Públicos de Tijuana. 2003. *Plan maestro de agua potable y saneamiento en los municipios de Tijuana y Playas de Rosarito*. February. Tijuana, B.C.: CESPT.

Comisión de Servicios de Agua del Estado de Baja California. 1994. *Plan estatal hidráulico*, 1994–2015. December. Mexicali: CSAEBC

Comisión Estatal del Agua de Baja California. 2003. *Programa estatal hidráulico 2003–2007*. September. Mexicali: CEABC.

Comisión Nacional del Agua. 2000. *Programa hidráulico de gran visión. Región I, Península de Baja California, síntesis básica*. February. Mexicali: CNA.

Cortez Lara, A., and M. R. García Acevedo. 1999. "The Lining of the All-American Canal: The Forgotten Voices." *Natural Resources Journal* 40(2): 261–279.

Lorey, D. E. 1999. *The U.S. Mexican Border in the Twentieth Century*. Wilmington, Del.: SR Books.

Román Calleros, J., and F. Bernal. 1995. "Diagnóstico general de Baja California. El agua que cae ... el agua que corre." In *Agua, Salud y Derechos Humanos*, Iván Restrepo, ed. México, D.F.: Comisión Nacional de Derechos Humanos.

Román Calleros, J. 1990. *Origen y desarrollo de dos áreas de riego*. Tijuana, B.C.: El Colegio de la Frontera Norte.

Secretaría de Recursos Hidráulicos. 1972. *Resumen del estudio geohidrológico del Valle de Mexicali, B.C. y Mesa Arenosa de San Luis, Sonora, México*. México, D.F.: SRH.

Tapia, M. A. 1989. "Situación y perspectivas del suministro de agua potable y desalojo de aguas negras en Tijuana." Pages 111–117 in *Agua y Desarrollo Regional*. Mexicali, B.C.: Colegio de Economistas de Baja California.

Unikel, L. 1976. *El desarrollo urbano de México: Balance y perspectivas*. México, D.F.: El Colegio de México.

Zenteno, R. M., and R. Cruz. 1992. "A geodemographic definition of the northern border of Mexico." Pages 29–41 in *Demographic Dynamics of the U.S.-Mexico Border*, J. R. Weeks and R. Ham-Chande, eds. El Paso, Tex.: University of Texas at El Paso.

VIII

The All-American Canal: Perspectives on the Possibility of Reaching a Bilateral Agreement

Stephen P. Mumme and Donna Lybecker

Introduction

Mexico's dispute with the United States over the water that seeps from the unlined All-American Canal (AAC) is one of the most intractable problems in U.S.-Mexican relations. This chapter draws on several analytical perspectives—those of international law, common-pool resource theory, and game theory—to identify possibilities for settling the dispute. Recent development of international law appears to reinforce Mexico's position that the canal lining project is an international issue that may be subject to litigation. Moreover, all three of these analytical perspectives, when applied to the AAC lining dispute, suggest the possibility of resolving the problem through issue linkage and tradeoffs that offer mutual benefits to both countries and reduce the zero-sum, loser pays effects on Mexico. Recent institutional developments at the bilateral level provide new opportunities for crafting a cooperative solution through issue linkage. Recent U.S. unilateral measures, however, limit certain trade-off options. Mexico may need to pursue dispute resolution at the international level to persuade the United States to negotiate the dispute.

Background

In the past 25 years, few transboundary water disputes along the U.S.-Mexican border have proven as intractable as the one involving the groundwater associated with the AAC. This partly earthen canal, which parallels the border with Mexico and tracks within a handful of miles of the international boundary for more than half of its 82-mile reach west of the Colorado River, is the sole fresh-water source for the Imperial Irrigation District (IID) and a partial source for Mexico's rich Mexicali Valley agricultural zone. As such it is a principal marker for the region and a potent symbol of the wealth and prosperity of one of the richest agricultural zones in the continental United States.

When it became fully operational in 1942, the AAC's proximity to the border resulted in steady percolation of water into the sandy substratum, which after time passed to Mexico via the underlying gradient (Waller 1992). This groundwater source replenished existing subterranean water stocks near the border on the Mexican side and became an important water source for the Mexicali Valley as well water was exploited by electrical pumps.[1] The importance of this unintended water source for the Mexicali Valley became apparent in the decade-long Salinity Crisis that erupted in 1961, causing Mexico to turn more heavily to groundwater as an alternative source of fresh water (Mumme 1988). When the Salinity Crisis was settled in 1973, the instrument of settlement, the International Boundary and Water Commission's (IBWC) Minute 242 (IBWC 1973), committed both countries to a diplomatic procedure of limiting withdrawals in the nearby San Luis agricultural zone and reporting any changes in the extraction of transboundary groundwater resources along the entire border with a prospect of eventually reaching a border-wide arrangement for resolving future groundwater disputes.

The settlement of the Salinity Crisis, however, drew U.S. attention to the potential for recapturing this resource. By 1980, competition for Colorado River water supplies in Southern California and pressure for water conservation in the large IID and nearby Coachella Irrigation District led to specific proposals for lining the AAC and its siphon, the Coachella Canal (Waller 1992). These proposals naturally alarmed Mexico, which believed it had established a beneficial use of these unclaimed waters and sought to negotiate the question (Waller 1992; Hayes 1991). The United States, dominated by powerful upstream interests, disagreed and proceeded unilaterally through the U.S. Department of the Interior's Bureau of Reclamation (BOR) to advance options for reclaiming lost AAC and Coachella Canal water. In 1988, Congress authorized the U.S. Secretary of the Interior to proceed with the lining project for both canals (Waller 1992). In 1994, a final Environmental Impact Statement, as required under terms of the U.S. National Environmental Policy Act, recommended that the United States construct a 23-mile, concrete-lined canal parallel to the AAC that would reclaim a total of 67,700 acre-feet (AF) of water annually (BOR 1995). This Environmental Impact Statement was reviewed in 1998 and 2002 and determined to have met all legal requirements (Dibble 2005).

Shortly after the initial 1988 decision to line the AAC, Mexico informally lodged a protest with the United States through the Comisión Internacional de Limitíes y Aguas (CILA), its section of IBWC, alleging potential injury to existing uses in the Mexicali Valley and pointing to the U.S. obligation arising from Minute 242 to consult with Mexico on any changes affecting groundwater use along the international boundary (Rohter 1989). Mexico's case centers on the issue of so-called acquired rights. The United States, in turn, denied any intent to alter the groundwater regime, claiming AAC seepage water was, in fact, previously allocated to the United States and its status remained unaffected by any conservation measures the United States should take (Rohter 1989).

These issues remain today. The U.S. government, despite protests in Mexico, has shown no intention of revising its decision to line the AAC. The United States and Mexico are at a diplomatic impasse on how to deal with the Mexican impacts while the AAC project itself has been delayed by funding issues involving the U.S. federal and California state governments. In the meantime, however, pressures on California's water supply and developments in the Colorado River basin have increased the stakes of any resolution at both the international and domestic levels. These developments are various but a short list would include the full operation of the Central Arizona Project, the 1998 San Diego-Imperial Irrigation District water conservation and transfer agreement, the 1999 Southern California Water Quantification Agreement, the 2000 California Colorado River Water Use Plan, and BOR's 1999 Interim Surplus Criteria Plan for the Colorado River.[2] Additionally, some water from the lining of the canal would go to the San Luis Rey Reservation for settlement of a decades-old legal fight (Dibble 2005). In Mexico, the privatization and transfer of water rights, the search for additional water supplies for Mexicali, and planning for a new Colorado-Tijuana aqueduct influence thinking on the use of Colorado River water supplies and the need to conserve existing sources. At the binational level, non-traditional water use concerns have arisen most intensely with respect to the Colorado River delta-estuary, attracting international environmental interest.

In 2002, the California Department of Water Resources and IID signed an agreement that provided $126 million to construct and line the 23-mile AAC replacement canal and set a completion date of 2006 (Department of Water Resources 2002). Owing to delays the project is now scheduled to begin by spring 2006 and be completed by December 2008 (Dibble 2005). At the international level, the United States and Mexico have formed a working team at IBWC to consider international aspects but with no deviation from original legal arguments on the merits of the situation. On the U.S. side, the official view is reflected in the Colorado River Water Use Plan, which acknowledges no obligation to Mexico but as a matter of "good will" encourages Mexico to consider conveying a part of its 1944 Water Treaty allotment to the Mexicali Valley through the newly lined AAC, thereby providing "a water quality benefit to Mexico" (Colorado River Board of California 2000). Mexico, through CILA, continues to object to unilateral measures in the matter of the lining project and has pressed the United States to continue discussing the impacts, technical aspects, and related water matters such as the management of Colorado River exceedances (flood waters and surge flows) that might figure into a compensatory and cooperative solution to the problem.[3]

While the two countries have thus far maintained an amicable posture toward each other in this dispute, the progress of events is undoubtedly prejudicial to Mexico, making this issue one of the most intractable problems in border groundwater management. To shed greater light on the options for resolving the dispute, this chapter proceeds by examining the case from three scholarly perspectives that may suggest areas of tractability that could be used in the search for a cooperative binational solution. The article begins by reviewing arguments in international prescriptive law that bear on the case. It then considers the problem from the perspective of common pool resource theory to better understand the structural elements of the case that may be conducive to a settlement. These structural arguments are followed with a look at the case from the perspective negotiating theory to identify certain options and leverage points that may improve the prospects for a cooperative solution. The conclusion summarizes the main findings from each of these perspectives and suggests which may be most promising as negotiators search for an accommodation in this case.

International Law and the All-American Canal Lining Dispute

One opportunity for a cooperative solution to the AAC lining dispute is found in international prescriptive law. Recent developments in international law have sought to extend principles of international law to the problem of transboundary groundwater management. Of course, in this instance, part of the dispute lies in whether or not AAC seepage water constitutes "groundwater" as such or remains a part of the surface water regime. Even so, since one of the disputing parties, Mexico, argues seepage should be treated as groundwater, it is worth considering the problem as if this were so. Moreover, as some prominent international legal scholars argue, more general rules also apply whether or not the seepage is considered groundwater.

An important set of arguments is presented by professor Albert E. Utton, who, until his death, was the editor of *Natural Resources Journal* and a keen student of international legal applications to transboundary resource disputes. In 1991, Utton specifically addressed the AAC problem, noting that several principles of international and domestic law should apply to any resolution of the dispute. He first noted that the principle of limited territorial sovereignty clearly applied to the 1944 Water Treaty that governs surface water allocation on the Colorado River (Utton 1991). The principle of limited territorial sovereignty, in turn, was fundamental to the rule of equitable apportionment adopted by the International Law Association (ILA) at Helsinki in 1966 and incorporated in solutions to many international disputes over the use of surface waters among co-riparians.

While the Helsinki rules do not formally address groundwater apportionment, as Utton observes, they do have limited application to groundwater (Utton 1991). In 1986, the ILA adopted a rule that incorporated aquifers that "receive water from surface waters of an international basin" as part of the larger river basin in question (Utton 1991). By this provision the Helsinki rules are extended to groundwater basins of the type found in the AAC situation. More recently, the United Nations' International Law Commission (ILC) reinforced this position in its International Convention on the Law of the Non-Navigational Uses of International Watercourses, a document completed and made available for signature in 1997.[4] In this case the ILC defined an international watercourse as "a system of surface waters and groundwaters constituting by virtue of their physical relationship a unitary whole and normally flowing into a common terminus" (McCaffrey 1996). The Convention's General Principles embrace the doctrine of equitable and reasonable apportionment found in the Helsinki rules, and also embraces the principle that nations must avoid "significant harm" in arriving at equitable solutions to international watercourse disputes. Legal theorist Stephen McCaffrey (1996) argues that in balancing the principles of equitable utilization and doing no significant harm, the convention favors a process based on inter-state discussion of all the factors involved that allow flexibility in the resolution of disputes.

These international legal principles, as articulated both by legal professionals and governments, would not apply directly to the AAC case unless the dispute advanced to an international dispute resolution forum like the International Court of Justice at the Hague (World Court) or arbitration under the Inter-American Arbitration Treaty (Hayes 1991). Even so, in the absence of a bilateral agreement clearly sustaining a U.S. right to AAC seepage, they provide general legitimation for treating the AAC dispute as an international, and not strictly domestic, problem and establish the basic principles that apply to any attempt to resolve the controversy at the international level.

As Utton (1991) comments, the directly applicable international law in the AAC case is the bilateral 1944 Water Treaty,[5] which is silent on the question of groundwater while allocating 1.5 million acre-feet of the Colorado River's surface water to Mexico. The treaty provides that "Mexico shall acquire no right beyond" this specified amount of surface water.[6] While this language seems definitive, the fact that treaty negotiators were cognizant that groundwater in Baja California was hydrologically connected to the Colorado River but silent as to the potential effects of Mexican use of those groundwaters on the hydrological system constitutes a major ambiguity that the treaty does not resolve in a mutually satisfactory manner (Utton 1991). Indeed, the treaty could be read as "requiring maintenance of the underground flow that existed at the time the treaty was made" (Kishel 1993). This being so, Utton argues, the international doctrine of prescription may apply. Under the rule of prescription, conferral of title of ownership may apply to the case of protracted possession and use of the resource where no original title or proprietary right can be found, or the original possession was wrongful, or the legitimate proprietor failed to assert his right (Utton 1991). Underscoring the U.S. potential liability should it curtail AAC seepage flows to Mexico is IBWC's Minute 242 (IBWC 1973), a protocol to the 1944 treaty that provides that the United States and Mexico "consult with each other prior to undertaking any development of either the surface or groundwater resources, or substantial modifications of present developments, in its own territory in the border area that might adversely affect the other country." While this provision requires only consultation, not consent (Hayes 1991), it has the further effect of linking groundwater to the 1944 treaty, thereby reinforcing the international aspect of the problem.

International law, then, provides a substantial basis for treating the AAC lining project as a legitimate international dispute to which international legal principles should apply. If the AAC dispute can be legitimately treated as an international dispute, and not simply as a domestic matter, as U.S. interests frequently assert, then Mexico may have recourse to international and multilateral dispute resolution mechanisms should the United States proceed with its plan to line the canal. Raising the dispute from the bilateral to the international level has risks for both parties that should create incentives favoring negotiation rather than confrontation.

Common Pool Resource Theory and the All-American Canal Lining Dispute

Yet another perspective on avenues for a cooperative solution of the AAC lining dispute may be drawn from the writing of common pool resource theorists. Common pool resources (CPR) theory aims to establish the institutional conditions that favor either cooperative or conflictive solutions to CPR-based disputes. One of the principal exponents of this school of thought, Elinor Ostrom (1990), identifies four factors conducive to cooperative solutions in CPR disputes:

- The resources are not depleted beyond recovery
- Resource conditions can be reliably determined
- The resource is sufficiently predicable
- Resource distribution is localized such that it can be evaluated and regulated

For the AAC, most of these conditions apply. However, the existence of different appropriation rules at the national and subnational levels, based on notions of divisibility and subtractability, complicate arrival at common management approaches. Binationally, the asymmetrical control of surface and groundwater resources in the Colorado River basin adds a strong disincentive to cooperation. National variations in appropriation rules constitute what Ostrom and others call a "constitutional" problem (Ostrom 1990; Burger and Gochfeld 1998), where different rules create different sets of expectations and engender uncertainty as to the application of the rules on either side of the border. This, in turn, adds an incentive to deplete the resource. The problem of constitutional uncertainty is amplified by national asymmetry of control that allows one community of users to deny others access to the resource.

CPR theorists argue that certain variables work as incentives to promote cooperation in settling common-pool disputes. These variables include, according to Burger and Gochfeld (1998):

- Embracing the opportunity for incremental change
- Developing a common frame of reference for the design of rules and regulations at the constitutional and collective action levels
- Increasing the quality and accessibility of information for the community of users, publicizing the mutual benefits and harms arising from current or status-quo rules and practices, as opposed to proposed rules and practices

To some degree, progress has been made along the border in meeting these conditions, particularly in the areas of developing better technical knowledge of groundwater basins and improved access to that information. The two countries remain far apart, however, in their construction of harms and benefits regarding the use of transboundary groundwater basins and in developing a common frame of reference on the rules and regulations that should apply to the use of the resource.

Viewed from the CPR perspective, the AAC lining dispute does not appear very amenable to solution. The quality of information available to the parties remains unequal because the two countries have been fairly reluctant to share the basic hydrographic information bearing on the dispute. The two nations also differ on the costs and benefits of the lining itself. Groundwater management practices differ substantially across the border, providing little basis for cooperation arising from a sense of mutual norms. And the unilateral character of the U.S. decision to recapture seepage from the AAC takes the form of a non-incremental, abrupt alteration of the status quo that is conflictive and not conducive to building bilateral commitments.

Even so, CPR theory does point to certain actions that could improve the prospects for a cooperative solution in the AAC lining dispute. First, CPR certainly lends support to the need to build a common and mutually acceptable database on the transboundary impacts of the lining project and related technical aspects of water use in the region that may be relevant in fashioning solutions to the dispute. Second, CPR suggests that efforts to publicize the adverse effects of the project while simultaneously advancing mutually beneficial options for resolving the dispute would help in reframing public perceptions of the problem. Third, CPR lends support to the value and importance of strengthening international legal norms applicable to the case, which occurred with the United Nations Convention on Non-Navigational Uses of International Watercourses. In this aspect, CPR points the way to the types of initiatives most likely to steer the conflicting parties toward a mutually acceptable settlement, if one is to be found.

Game Theory Applications to the AAC Lining Dispute

Additional insights on the prospects for resolving the AAC dispute in a cooperative fashion have been drawn from game theory. In a very useful article, Frisvold and Caswell (2000) argue that the AAC lining dispute, as it presently stands, can be modeled as a two-player non-cooperative/prisoner's dilemma game with unidirectional externalities. As they observe, victim-pays regimes should be regarded as unsatisfactory on several grounds, including the fact that income-poor downstream parties may be unable to offer side payments to prevent upstream parties "from polluting or diverting transboundary waters" (Frisvold and Caswell 2000). In the AAC case, Mexico is victimized by U.S. unilateral actions upstream.

One option available to downstream parties seeking to avoid side payments or avoid suffering the alternative of accepting negative externalities is issue linkage. In the language of game theory, issue linkage is expressed as an interconnected game or two-level game in which outcomes at one level are contingent upon outcomes at another. As Frisvold and Caswell (2000) put it, "this allows for equilibrium solutions not attainable in isolated games that may yield higher joint payoffs." They note that the logic of this type of game structure suggests the utility of linking negotiations on the canal lining project to other water-related concerns of binational interest. Water transfers, water conservation, pollution control and mitigation, and eco-system protection would all fit into this category. Linkage may also be made to other issues of binational import, though this might introduce greater risks in the incentive structure for binational negotiations.

In the AAC lining case, a wide range of linkage options certainly present themselves. Kishel (1993) lists the transfer of fresh water produced from the Yuma Desalination Plant by pipeline to Mexicali, groundwater banking, U.S. assistance with water conservation in the Mexicali Valley, effluent exchange opportunities, and the environmental benefits to the United States to be derived from any of the above, as potential negotiating points (Kishel 1993). To this could be added improved water conveyance through the newly lined canal, assistance with the proposed Mexicali-Tijuana-San Diego aqueduct, transfers (including effluents) to the Colorado River delta ecosystem, links to potential benefits in settling U.S. tribal nation water rights claims, endangered species claims, and other claims related to U.S. and California administrative solutions to meet California's federally mandated water conservation targets by 2015.

In sum, game theorists point in the direction of a negotiated solution to the AAC based on a strategy of issue linkage. When combined with the advice from CPR theory and international legal scholars a clear preference can be seen for efforts at steady institution building (CPR theory), negotiated solutions (CPR and game theory) and, failing that, for international litigation or arbitration (international law). The next section offers suggestions toward a practical course of action consistent with these perspectives on the dispute.

Toward a Negotiated Solution to the All-American Canal Lining Project

While the specific features of the AAC lining dispute are certainly not optimal for conflict resolution, all three perspectives outlined above point to some aspects that could be used in a cooperative solution. From an international law perspective, there is sufficient ambiguity at the level of interpreting the 1944 Water Treaty and Minute 242 to argue that Mexico does have a basis for raising the issue to the level of international dispute settlement mechanisms should the United States persist in implementing a unilateral outcome. This is true regardless of whether the water is treated as surface or groundwater—and it bears mentioning that some agreement on the area's transboundary groundwater is necessary regardless of the specific problem of seepage influx from the AAC. But a legal solution, much like the unilateral outcome itself, amounts to a zero-sum, loser pays, sub-optimal solution that entails high risks for both parties. While in a narrow technical sense justice might be done by litigating the dispute, a litigated outcome is hardly the best mechanism for advancing binational cooperation or achieving a management outcome consistent with the emerging norms of sustainable development of the region's water resources.

The three perspectives also point to opportunities for achieving a negotiated solution. All three perspectives, for instance, point to the importance of linking related issues in ways that generate benefits for both countries. And all three encourage the provision of better information and transparency in negotiating any solution.

Short of a litigated outcome, then, a number of options exist for linking related water management issues that can contribute to the sustainable development of water resources in the region. In point of fact, two of these options have been proposed by representatives of the IBWC's national sections. Within the context of Minute 242, the IBWC's national sections have informally agreed, for instance, to study the potential technical and ecological impacts of groundwater reduction related to AAC lining and to consider any cooperative measures that could be taken to mitigate adverse effects.[7] IBWC has also offered to allow some delivery of Mexican treaty water through the lined canal (Colorado River Board of California 2000). While these measures are not sufficient to meet Mexico's demands and expectations for a cooperative solution to the dispute, they should be treated as a useful expression of international comity that effectively recognize the issue as a critical point of binational concern and try to address that concern.

Moving beyond these limited measures most likely requires some linkage among related issues of binational concern. Following Kishel's (1993) lead and updating his argument for a constructive regional water planning approach to settling the AAC dispute, several options certainly appear to have merit in developing a linked strategy of tradeoffs or compensation in a negotiated solution to the dispute. One such option links a compromise on AAC seepage water to a proposed binational aqueduct linking the Colorado River to Tijuana-Ensenada and San Diego on the Pacific Coast. The two countries are seriously concerned with achieving greater capacity in delivering Colorado River water to these areas (IBWC 2000b). In 1999 IBWC concluded Minute 301, aimed at developing strategies for constructing such an aqueduct (IBWC 1999). One of the leading alternatives calls for building a parallel line to the existing Mexicali-Tijuana aqueduct and transferring San Diego water through that conveyance (San Diego Dialogue 2001). Such a plan would significantly reduce engineering and construction costs to the United States. Mexico's cooperation could be linked to U.S. proposals to replace all or part of the 67,000 AF of lost seepage to Mexico, or to some form of monetary compensation in support of Mexican infrastructure development in various forms, including increasing conservation in irrigated agriculture or subsidizing aqueduct construction costs.

Tradeoffs aimed at achieving a sustainable solution to the water requirements of the Colorado River Delta are also an opportunity for a cooperative solution. Upstream conservation measures in the United States, particularly BOR's Interim Surplus Criteria master plan for cushioning California's transition to a reduced volume of Colorado River water over a period of 15 years, will diminish so-called pulse flows, possibly one of the few sources of water sustaining the delta's ecosystem (Culp 2000). The delta's plight has aroused the concerns of international environmental groups who fear the potential loss of wildlife habitat and adverse effects on aquatic species in the upper Gulf of California.[8] This mobilization of environmental concern has, in turn, led to a search for alternate water sources that might compensate for reduced Colorado River flows to the delta (Pitt, el al. 2000). Environmentalists on both sides of the border say that without compensation, lining the canal threatens wetlands and important border habitat (Dibble 2005).

The problem of delta conservation is linked to the general hydrology of the lower Colorado River, including groundwater flows in the region. This fact provides a practical trace to the AAC seepage dispute. The two countries agreed in December 2000 to a framework agreement for the study of delta conservation options (IBWC 2000b), placing the issue within the broader language of sustainable development. Current options under consideration include mixing agricultural wastewater with treated wastewater as a means of replenishing pulse flows, conserving Mexicali Valley agricultural water and transferring savings to the delta, purchasing and retiring water rights in Mexico or the United States and dedicating these to delta conservation, or some combination of these (Pitt, et al. 2000). In the case of the delta, what is certain is that monetary compensation alone is incompatible with a solution; some additional water is necessary to replace pulse flows and save the ecosystem. This, in turn, suggests the possibility of trading reclaimed seepage water for monetary or other compensation and using the water either directly or indirectly (through a system of water trades and transfers in the Mexicali Valley) to augment delta water stocks (Pitt, et al. 2000). Such a solution has the advantage of directly connecting the use of AAC seepage water to a recognized binational objective of substantial benefit to both countries.

Another option is found in current Mexican proposals to develop desalination facilities to augment area water supplies. As Kishel (1993) argued more than a decade ago, reclamation of water through desalinization remains an option for dealing with water shortages in the region (Kishel 1993). However, piping water to Mexicali from the long delayed Yuma Desalinization Plant, as Kishel (1993) suggested, is now a less-viable option. The plant became operational in 1992 but after a year was shut down due to cost (Clinton, et al. 2001). BOR has recently indicated it may resume operations, but if resumed, any processed water discharged to Mexico will count against its treaty quota to free up additional water for U.S. clients (Pitt, et al. 2000). In the context of the new Interim Surplus Criteria management plans, changing these practices would be quite controversial. However, the fact that Mexico's Comisión Nacional del Agua (in English National Water Commission) is now beginning to look at long-term desalinization solutions for Baja California's urban centers suggests a possible solution whereby the United States helps finance Mexican desalinization through the North American Development Bank (NADBank), the Interamerican Development Bank, or World Bank, with the clear understanding that such financing is in compensation for Mexico's lost seepage water and that in some fashion the enhanced water supply would be linked to the binational effort to restore the delta ecosystem. The authorization of NADBank to lend to border projects with a clear environmental component makes this a serious option in dealing with the seepage problem (NADBank 2000).

Yet another option is relying on market mechanisms to deal with water availability problems. Water markets and water banking have been suggested as options for augmenting delta water supplies and are bearing serious examination by major environmental organizations involved in delta conservation efforts (Pitt, et al. 2000). A market solution to the AAC dispute would probably entail purchasing a volume of water equal to or in some proportion to the volume of reclaimed seepage water and making that available to Mexico or, conversely, simply compensating Mexico directly with a cash settlement for the AAC water withheld by the United States. Both these alternatives are highly controversial as stand-alone policies. They take the form of zero-sum solutions (only one party gets water) that are likely to raise both substantive and symbolic objections in Mexico and the United States. At the very least, any market-based solution would need to be linked to other offsetting benefits to make the deal more attractive to one or both parties.

Banking surplus flows of Colorado River water as groundwater, as Kishel (1993) suggested, has been built into BOR's Interim Surplus Criteria Plan but in a manner adverse to Mexico and the delta rather than one conducive to the settlement of the AAC dispute—the storage is upstream in the United States rather than in Mexico (Colorado River Board of California 2000). Thus, the solution he envisioned seems less viable today. Even so, particularly where wastewater reclamation is concerned, there may be some options for banking reclaimed water in a manner that permits greater flexibility in extraction and use in the region. The potential sources of reclaimed wastewater in the region are various and include agricultural sources, urban sewage and drainage in both countries' border cities, and as Kishel (1993) observed, possibly even sources as distant as Riverside, California, and the Los Angeles metropolitan area should the financing become available to support the necessary infrastructure.

Other linkage options certainly exist and could be explored. As Frisvold and Caswell (2000) argue, the Border XXI and Border 2012 Programs that operates under the authority of the 1983 La Paz

Agreement provides an institutional framework for exploring a range of mutual interests involving water-related and other environmental concerns that could be used to identify and prioritize potential policy links that are advantageous for both countries and contribute to border sustainable development objectives. The same can be said for the Border Environment Cooperation Commission and NADBank, whose mandates have recently broadened beyond water infrastructure projects to embrace environmental protection and natural resources conservation (NADBank 2000). It is certainly true that the maturation of these border institutions has strengthened the institutional bases of mutual understanding and communication that common pool resource theorists deem essential to achieving negotiated solutions to problems like the AAC lining dispute.

Conclusion

A number of linked solutions are available for consideration should the two countries choose to seek a cooperative solution to the AAC lining dispute. As others have argued, it makes sense to look at this dispute through the lens of greater economic integration and mutual cooperation rather than through the narrow slit of issue-specific, zero-sum, loser-pays outcomes. Unfortunately, bilateral discussion of the AAC lining dispute has thus far framed the issue in zero-sum terms, if only by default. Apart from the limited commitments made at the level of IBWC, there is little evidence of a willingness in the United States to consider Mexico's concerns. Even the Environmental Impact Statement on the canal lining project, completed in 1994, did little to incorporate Mexico's concerns in its final recommendations, proceeding on the assumption the legal and international issues were essentially settled (Jones, et al. 1997). They are not. Both California and BOR have proceeded unilaterally to take measures that have the effect of foreclosing certain options for settling the dispute and this, in turn, strengthens the prospect of a victim-pays outcome to the controversy.

As construction nears on the canal lining project, now slated for completion in 2008, it appears increasingly necessary that Mexico take its case to the World Court to receive a fair hearing on the issue. Where other mechanisms fail, litigation remains a reasonable option in the quest for an equitable resolution to the problem. It bears noting that part of the difficulty in pressing for a negotiated solution is the culture of water ownership in the United States, where water rights are a critical part of the private property regime and jealously defended, with disputes over allocation typically resolved in court. Nowhere is this more true than in the Colorado River basin where over-allocation of water supplies, rapid urbanization, and the emergence of non-traditional claims on water supplies ranging from the quantification of Native American water rights to environmental conservation have led to all-out competition for the basin's water. As others have argued, international water disputes in which the predominant or upstream party is accustomed to litigating its domestic disputes over water allocation and also has the capability of unilaterally withholding the resource from the disputing party, are the very cases that are most likely to result in international litigation.

In the AAC lining dispute, should Mexico elect to press for international dispute resolution, there is some question as to whether the United States could exempt itself from the World Court's jurisdiction under the Connolly Amendment. The Connolly rule allows the United States to exclude itself from the court's compulsory jurisdiction in those cases of "disputes with regard to matters which are essentially within the domestic jurisdiction of the United States of America as determined by the United States of America" (Von Glahn 1992). Even if the court could claim compulsory jurisdiction, Mexico would run the risk of vacating its claim to seepage water should the court find in favor of U.S. arguments. However, the heightened risk to both parties of losing important values in a litigated outcome might persuade the party resisting negotiation to seek a negotiated settlement. In the absence of alternatives, a strategy of using litigation to force the United States to the bargaining table may be Mexico's only option, though Mexico, as it did in the 1960's in the Salinity Crisis,[9] will need to weigh the risk of confrontation against its broader bilateral objectives. Against this gloomy prospect, the good news is that many linkage options exist that could be explored in the search for a negotiated solution and that recent institutional developments at the bilateral level should facilitate the search for mutually beneficial tradeoffs.

Endnotes

[1] In fact, Colorado River water, originally conveyed to the Mexicali and Imperial valleys through Mexico's Alamo Canal, has seeped continuously into the Mexicali Valley. After the AAC came fully online in 1942, Mexico, by contract with IID, continued to receive its portion of Colorado River water through the Alamo Canal by means of an unlined diversion from the AAC. This practice continued until the 1950 completion of Morelos Dam, Mexico's diversion dam and the lowest dam on the Colorado River. After 1950, Mexico continued to receive a small part of its Colorado River water allotment through the AAC under agreement with the U.S. Bureau of Reclamation and IID. This fact, coupled with IID's long ownership and operation of the Alamo Canal after 1916, greatly complicates the question of assigning responsibility and ownership of the seepage. See, Ward (2003) and Enriquez-Coyro (1975).

[2] A discussion of these various developments may be found in Colorado River Board of California 2000.

[3] This is according to an interview with Arturo Herrera Solís of CILA on May 30, 2001.

[4] This can be found in volume 36 of *International Legal Materials.*

[5] Officially called the *Treaty Respecting Utilitization of the Colorado and Tijuana Rivers and Rio Grande*, it was signed February 3, 1944 between the United States and Mexico.

[6] This provision appears in Article 10, Section b.

[7] This is also according to Arturo Herrera Solís.

[8] For a list of major groups involved in delta restoration see Varady, et al. (2001).

[9] Declassified State Department documents available in the U.S. National Archives show that State Department and Interior Department officials very much feared that Mexico would take its case to the World Court in 1969 and 1970. These reckonings indicate that under threat of litigation, the United States preferred to negotiate an outcome to the problem of salinity. See U.S. National Archives, Record Group 59, Subject-Numerical Files 1967–1969, Box 2347, Pol 33-1 Mexico-United States. It is worth noting that Mexico's threat to take the salinity problem to the World Court was taken very seriously as early as 1964. See Thomas C. Mann,

Assistant Secretary of State, letter to Steward Udall, Secretary of the Interior, April 30, 1964. Washington, D.C.: Department of the Interior, Identification No. 2884. Classified Confidential. Stewart Udall Papers, Special Collections, Library, University of Arizona.

References

1944. *Treaty Respecting Utilitization of the Colorado and Tijuana Rivers and Rio Grande*. 3 February.

1997. "Convention on the Law of the Non-Navigational Uses of International Watercourses." http://www.un.org/law/ilc/texts/nonnav.htm.

Bureau of Reclamation. 1995. *Final Environmental Impact Statement/Final Environmental Impact Report: All-American Canal Lining Project, Imperial County, California*. Washington, D.C.: U.S. Department of the Interior.

Burger, J., and M. Gochfeld. 1998. "The Tragedy of the Commons 30 Years Later." *Environment* 40.

Clinton, M., M. Hathaway McKeith, J. Clark, P. Cunningham, D. H. Getches, J. L. Lopezgamez, L. O. Martínez Morales, B. Bogada, J. Palafox, and C. Valdés-Casillas. 2001. *Immediate Options for Augmenting Water Flows to the Colorado River Delta in Mexico*. Report prepared for the David and Lucille Packard Foundation. Boulder, Colo.: Clark Consulting, Inc.

Colorado River Board of California. 2000. *California's Colorado River Water Use Plan (draft version)*. Sacramento, Calif.: Colorado River Board of California, The Resources Agency, State of California.

Culp, P. W. 2000. *Restoring the Colorado Delta with the Limits of the Law of the River*. Tucson, Ariz.: Udall Center for Studies in Public Policy

Department of Water Resources. 2002. "All-American Canal Lining Agreement Signed." News Release. Sacramento, Calif.: Department of Natural Resources.

Dibble, S. 2005. "Wetlands become a focus in debate over canal lining." *San Diego Union-Tribune* 6 January.

Enriquez-Coyro, E. 1975. *El Tratado entre México y los Estados Unidos de América sobre Ríos Internacionales, Vol. I.* México, D.F.: Facultad de Ciencias Políticas y Sociales, Universidad Autónoma de México.

Frisvold, G. B., and M. F. Caswell. 2000. "Transboundary Water Management: Game-theoretic Lessons for Projects on the U.S.-Mexico Border." *Agricultural Economics* 24: 101–111.

Hayes, D. L. 1991. "The All-American Canal Lining Project: A Catalyst for Rational and Comprehensive Groundwater Management on the United States-Mexico Border." *Natural Resources Journal* 31(4): 806.

International Boundary and Water Commission. 1973. *Minute 242: Permanent and Definitive Solution to the International Problem of the Salinity of the Colorado River.* http://www.ibwc.state.gov/Files/Minutes/Min242.pdf.

International Boundary and Water Commission. 1999. *Minute 301: Joint Colorado River Water Conveyance Planning Level Study for the San Diego, California–Tijuana, Baja California Region.* http://www.ibwc.state.gov/Files/Minutes/Min301.pdf.

International Boundary and Water Commission. 2000a. *1999 Annual Report.* El Paso: IBWC. http://.

International Boundary and Water Commission. 2000b. *Minute 306: Conceptual Framework for United States-Mexico Studies for Future Recommendations Concerning the Riparian and Estuarine Ecology of the Limitrophe Section of the Colorado River and its Associated Delta.* http://www.ibwc.state.gov/Files/Minutes/Min306.pdf

Jones, L. C., P. Duncan, and S. P. Mumme. 1997. "Assessing Transboundary Environmental Impacts on the U.S.-Mexican and U.S.-Canadian Borders." *Journal of Borderlands Studies* 12(1, 2): 81.

Kishel, J. 1993. "Lining the All-American Canal: Legal Problems and Physical Solutions." *Natural Resources Journal* 33(3): 697.

McCaffrey, S. C. 1996. "An Assessment of the Work of the International Law Commission." *Natural Resources Journal* 36(2): 304.

Mumme, S. P. 1988. *Apportioning Groundwater Beneath the U.S.-Mexico Border* Research Report Series 45. La Jolla, Calif.: Center for U.S.-Mexican Studies.

North American Development Bank. 2000. *Board Resolution 2000-10, Expansion of NADB Sectors of Activity.* 16 November. San Antonio, Tex.: NADBank.

Ostrom, E. 1990. *Governing the Commons.* Cambridge: Cambridge University Press.

Pitt, J., D. F. Luecke, M. J. Cohen, E. P. Glenn, and C. Valdez Castillo. 2000. "Two Nations, One River: Managing Ecosystem Conservation in the Colorado River Delta." *Natural Resources Journal* 40(4).

San Diego Dialogue. 2001. "The Water Challenge: Shrinking Supplies and Growing Demand." *San Diego Dialogue Report* 4(10): 1.

Rohter, L. 1989. "Canal Project Sets Off U.S.-Mexico Clash Over Water for Border Regions." *New York Times* 2 October.

Utton, A. E. 1991. "The Transfer of Water from an International Border Region: A Tale of Six Cities and the All American Canal." *North Carolina Journal of International Law and Commercial Litigation* 16(3): 480–481.

Varady, R. G., K. B. Hankins, A. Kaus, E. Young, and R. Merideth. 2001. "...to the Sea of Cortes: Nature, Water, Culture and the Livelihood of the Colorado River Basin and Delta—An Overview of Issues, Policies, and Approaches to Environmental Restoration." *Journal of Arid Environments* 49(1): 198–199.

Von Glahn, G. 1992. *Law Among Nations*, 6th ed. New York: MacMillan Publishing Co.

Waller, T. 1992. "Southern California Water Politics and U.S.-Mexican Relations: Lining the All-American Canal." *Journal of Borderlands Studies* 7(2).

Ward, E. 2003. *Border Oasis.* Tucson, Ariz.: University of Arizona Press.

IX

Opposing Approaches to Managing Shared Water Resources: The Lining of the All-American Canal and the Valley of Mexicali—Static Market Equilibrium or Nash Equilibrium?

Alfonso Andrés Cortez Lara

The region composed of the bordering states of California, Arizona, Baja California, and Sonora has an annual average water volume of 4.83 million acre-feet (MAF), which is supplied by the surface flows of the Colorado River and groundwater extracted principally on the Mexican side. The majority of this water is used to irrigate 450,000 hectares (ha) (1.1 million acres) of agricultural fields on both sides of the border. This water also supplies cities, including those on the coast such as Tijuana in Baja California and San Diego in California (Cortez, et al. 2002).

Despite the perceived abundant availability, growing competition for water, not only in this region but in the Lower Colorado Basin,[1] has led to efforts to decrease losses and optimize use through supply systems that would increase availability and meet demand, particularly in urban areas. The proposal to line the All-American Canal (AAC) is among the principal projects U.S. authorities are relying on to recover water losses.

The AAC conveys water from the Imperial Dam, 40 kilometers (km) (25 miles) upstream from the point of delivery, to Mexico at Morelos Dam, located in Algodones, Baja California. The AAC is the only source irrigating 198,296 ha (490,000 acres) and supplying water to 154,362 residents in Imperial County, California.[2] An annual volume of nearly 360,000 acre-feet per year (AF/y) flows through the AAC and U.S. government estimates indicate that seepage losses total as much as 91,610 AF/y, of which 10,540 AF/y flow north and 81,070 AF/y flow south. This water enters Mexicali's Irrigation District 014 as groundwater, where 56,750 AF/y are extracted from federally or privately owned wells and the remaining 24,320 AF/y are extracted by runoff collection at the La Mesa Drain.

The proposed lined AAC, which would run parallel to the existing canal, would be 37 km (23 miles) long and would begin 2.5 km (1.6 miles) west of Pilot Knob Peak. It would travel through the plateau of the same name, the Algodones Dunes, East Mesa, and end at Drop 3. The lining is expected to result in the conservation of 67,700 AF/y (DOI 2002).

Until very recently, the implementation of this project had been delayed by legal issues within the United States. With the enactment of Public Law 100-675 (U.S. Congress 1998), though, Congress authorized the U.S. Department of the Interior to move forward with the project.

However, international water regulations to which both countries are party seem to prohibit the project unless both the United States and Mexico agree to it, as set forth in the International Boundary and Water Commission's (IBWC) Minute 242, signed in August 1973. Specifically, it reads:

> With the objective of avoiding future problems, the United States and Mexico shall consult with each other prior to undertaking any new development of either the surface or the groundwater resources, or undertaking substantial modifications of present developments, in its own territory in the border area that might adversely affect the other country.

By carrying out this project the United States would be violating what was stipulated in this Minute and would also potentially cause adverse effects.

Water-Availability Trends in the Area of Influence

Studies carried out by Mexico's Comisión Nacional del Agua (CNA, in English National Water Commission), and in particular those conducted in 2000, indicate that lining the AAC would reduce water availability by as much as 81,070 AF/y. Of that water, 56,750 AF/y would have recharged the aquifer underlying the Mexicali Valley and 24,320 AF/y would have reached the La Mesa Drain (at its starting point, this drain runs on the Mexican side parallel to the AAC).

CNA also reports that all 328 wells—from which 3.16 MAF/y are currently extracted and used to irrigate 32,200 ha of agricultural fields—would be negatively affected to some degree or another. Negative effects would also be felt in the 1,200 ha currently irrigated with water collected at the La Mesa Drain. The total area facing potential affects by lining the AAC is estimated to be 33,400 ha largely located in the northeastern portion of the Mexicali Valley. Historically in this area a variety of crops have been grown, such as cotton (14,300 ha), wheat (8,400 ha), alfalfa (3,500 ha), vegetables (700 ha), and other smaller-scale crops (6,500 ha).[3] Another study indicates a reduction in the water table that would lead to a 15% increase in power and electric energy demand (Navarro 1998).

There will also be potential direct and indirect negative effects due to volume-quality ratios in the groundwater of the area of influence. Other authors whose estimates are more conservative suggest the direct effects would be felt in only 190 wells serving 19,000 ha of irrigated agricultural land and 800 ha irrigated by water from the La Mesa Drain. The difference in surface area is due to how gradually the effect spreads. It should be stressed that most agree that lining the AAC would cause direct short- and medium-term negative effects on the wells surrounding the area of influence of the AAC, and in the long-term, on the entire Mexicali Valley.[4]

Salinity Issues

The change in salinity is an issue directly related to reductions in available water volumes. According to CNA, in the area of influence of the AAC under current conditions salinity in the aquifer increases by 21.9 parts per million per year (ppm/y), and by 20.6 ppm/y near the old wells. Should the AAC be lined, total salinity in the aquifer would increase 5.7 times the expected norm during the first year. In subsequent years it would stabilize and exhibit an increase of 23.5 ppm/y. This suggests a quick water salinization process. Among the negative effects of increased water salinity on agriculture is a decrease in production of sensitive crops, which are indeed the most profitable.

Social Issues

These shared, fully-allocated crossborder waters further complicate an already difficult situation. The arid nature of the region, population growth trends, and economic activity dynamics make the fair and integrated management and distribution of water resources more complex, not only between the two countries, but also among its users—manufacturing sectors and population sectors (Cortez and Whiteford 1996).

Economic activity in the Imperial-Mexicali border region is predominantly agricultural, with most of the water (between 85% and 90%) used to irrigate crops such as cotton, vegetables, wheat, fruits, alfalfa, and grasses. Alfalfa and grasses alone use 54% and 15%, respectively, of the total amount of water available for agriculture in the region. This, combined with the steep increase in water demand for urban uses, has exerted additional pressure on agencies and manufacturing sectors, which have greater volumes of water available.

This border region includes the municipalities of Mexicali, Tecate, and Tijuana on the Mexican side, and the counties of Imperial and San Diego on the U.S. side. Together they make up the most populated area along the international border between Mexico and the United States. Mexicali, on the east side of the Baja California border, has an urban population of 647,950 and a scattered rural population that includes 214,545 in 294 villages.[5] The lining of the AAC would, in the short term, directly affect approximately 1,010 families, 950 of whom use water from the wells and 60 of whom benefit from runoff from the La Mesa Drain.

Impacts of Lining the AAC for Mexicali Valley

Several studies conducted by CNA consider the aquifers on both the Mexican and U.S. sides part of a single hydrogeological unit. The United States, though, conducted technical studies in 1988, 1994, and 2002 without any real participation from Mexican counterparts.[6]

The information collected by the IBWC on existing piezometric conditions in the Imperial and Yuma Valleys provides insight into their evolution since 1911. The static-level configuration analysis for different dates states that the construction of irrigation in the United States resulted in a change in the original groundwater flows on both sides of the border, and that some of the main "positive" effects of the AAC are the feeding of aquifers on the Mexican side, brought about by infiltration, and the formation of a piezometric dome at Yuma Mesa due to irrigation water infiltration.

The hydrogeological study performed in 1972 by CNA indicated that the Mexicali Valley aquifer (not taking into account Mesa Arenosa de San Luis) has an average annual recharge of 567,490 AF/y, divided as follows:

- 405,350 AF/y from vertical feed by infiltration from distribution canals in the Mexicali Valley irrigation area
- 121,600 AF/y from underground flows on the California-Baja California and Arizona-Baja California borders (including the AAC, whose underground flow also has a predominant vertical component)
- 40,540 AF/y from Mesa de San Luis Río Colorado in Sonora

Additionally, derived from the rehabilitation of Irrigation District 014, the La Mesa Drain was built with the initial purpose of intercepting infiltrated waters from the AAC to keep the elevated watertable levels that prevailed in the area—particularly in surrounding towns such as Algodones—under control.

With regard to water quality, several studies on the dissolved salt balance at the La Mesa Drain show it is made up of a 1,000 liter per second inflow from the AAC with an average salinity of 900 ppm, plus another similar volume from irrigation returns from the Mexicali Valley with a median salinity of 3,500 ppm. The greatest inflow of low-salinity water comes from the first section of the drain because this area has the highest permeability in the aquifer and is where infiltration from the AAC takes place (CNA 1991).

This information shows various qualitative and quantitative factors of the potential affects on the Mexican portion of the Mexicali Valley over different time periods, should the construction of the lined canal go forward. Despite the AAC issue having been discussed intensely since 1988, there is still no clear solution to "the new controversy," at least not a joint one. This is because, on the one hand, the U.S. government declares its intention to line the AAC and says it complied with public comment requirements under the 1944 Water Treaty, and on the other, the Mexican government simply expresses its disagreement, which achieves nothing in real terms. Meanwhile, the risk remains and the negative effects would, according to the available studies, be felt mainly in Mexican territory[7].

Therefore, it is worthwhile to analyze scenarios that could lead to successful negotiations between both countries to share this water, which is in high demand. It is not about holding back urban development or allowing agriculture to continue using the great majority of the region's available water inefficiently—one cannot hold back the development of one sector at the expense of another, nor maximize the growth of one region while minimizing the other.

Alternatives for Improving Negotiation Strategies and Binational Management of Water: Static Market Equilibrium or Nash Equilibrium?

Reaching viable alternatives lies not in technical solutions but in negotiation, conflict management, and "game" strategy, as well as with the role of and benefit for the players. The Game Theory perspective was introduced by Mumme and Lybecker in the previous chapter. The authors see that theoretical framework as one of the potential alternatives that might help to reach a bilateral solution to the AAC conflict. The authors, based on the findings of Frisvold and Caswell (2000), argue that in order to overcome the inherent constraints of this approach (non-cooperative choice game), issue linkages should be considered in the scenario.

In addition to the ideas developed by Mumme and Lybecker in this volume, this chapter explores two theoretical concepts that illustrate solutions to a complex situation: Sharing water when there is competition over this priority resource, which is strategic to regional development but also scarce, dwindling in supply, and increasingly in demand from various sectors in two countries.

What is the correct approach to analyze the All-American Canal conflict? Is it a Static Market Equilibrium—defined as the process that "optimizes" the allocation and use of resources, in this case, water? Or is it a Nash Equilibrium—a strategy that seeks a shared benefit, previously discussed and reached by consensus? Which ought to be the basis of negotiation to resolve this difference, Adam Smith's Price Theory[8] or the Equilibrium Theory of John Forbes Nash?[9]

Smith's is the predominant approach upon which water management in the United States, and particularly in the Lower Colorado River Basin, is based. This chapter argues that this theory has led to conflict among the different water users when attempting to achieve social efficiency. Although this mechanism works adequately on the U.S. side, thanks to its organizational conditions and available infrastructure, and especially the nature of its legal framework with regard to water property rights, it would be inappropriate for Mexico to enter into the water markets arena today, mainly because of the characteristics that property rights should possess. They should be:

- Clearly defined
- Reasonably complete
- Certain
- Transferable

The water-rights trade model based on Smith's theory generally states that, while buyers and sellers seek benefits for themselves or their group, the "invisible hand" of the market will distribute the goods efficiently. However, there are plenty of examples in which this "invisible hand" does not efficiently distribute resources and which this theory does not explain. One example is the California water markets and their resulting crossborder effects on water supply, as have been seen in recent years in the Lower Colorado Basin and the Imperial-Mexicali region. This development model leaves one of the parties—the Mexican agricultural sector—unprotected.

Thus, continuing to base crossborder water management (including water-saving and optimization projects, such as the AAC's lining) on Smith's theory would lead to marked conflicts when sharing a common resource like water (Ingram 1992). Therefore, it would be difficult to make predictions of the results of bilateral, multisectorial negotiations. In this context, a posture derived from a weighted negotiation could provide favorable solutions for all the stakeholders of these shared—and in fact over-allocated—waters of the Lower Colorado and/or for specific issues, such as in the case of the AAC lining project.

Nash's Equilibrium Theory and its derived Game Theory, though, fit the particular case of bilateral relations or conflicts, such as in the AAC lining. Essentially, it proposes the idea that every "game" (conflict or negotiation) of an economic nature has a point of equilibrium through which the "players" (political, financial, societal, and other stakeholders) can ensure ending up in a better position than before, even when it is not the ideal position for which they would have wished.

According to Nash, market system inefficiencies are due to the fact that the parties in a negotiation tend to look for the point where the decision or agreement reaches the optimum solution in the system, even if it is not beneficial or the most efficient for all players. Therefore, proposing common benefit strategies for dealing with a common resource like water could be a more appropriate approach.

The theoretical alternative suggested by Nash improves the situation when compared with that of Smith. However, Nash's alternative, which considers the dominant non-cooperative choice, would still lead the participants to a low level equilibrium that is not attractive to them. Nash equilibrium is defined as where no player can improve their position by acting alone if others stick to their positions (Schmid 2004). This has been described as a social trap (Platt 1973).

Nash's theory could be considered a theoretical basis on which to analyze issues related to shared water resources. Addressing crossborder water conflicts in a binational context, such as the AAC lining project, by using Nash's proposal could lead in new and better directions because it would be possible to ensure, with great accuracy, that all parties involved come out ahead, instead of just the one party that controls or dominates, as occurs in the market system.

Considering that the water seeping from the AAC toward Mexico could, on an institutional level, be categorized as an asset with a high exclusion cost, the Theory of Equilibrium, in which no player or participant can improve his position in the negotiation by acting unilaterally or individually, suggests that the best solution could be obtained by acting collectively and promoting the values of trust and cooperation with others (Schmid 2004). The issue is that trust is learned over time through successful transactions (human interrelations), including the learning of legal and behavioral norms. With regard to legal norms, which should be based on such interrelations, there are limited precepts. Although the international groundwater regulations between the United States and Mexico, as set forth in Minute 242, state that there must be "coordination for extraction and also consultation and exchange of information on the aquifer," it is precisely these elements of cooperation that have not been achieved. This gap in cooperative behavior could be closed using Nash's Equilibrium model.

The Nash's equilibrium model could be improved with Schmid's suggestion. Schmid states that the only way out of the social trap (do not cooperate) is by fostering trust and being confident in the cooperation of others. This can be achieved through intensive and permanent dialogue and repeated games between the parties involved—that is, through several negotiations sessions.

An in-depth review of available data (Navarro 1998; Cortez Lara 1999; Cortez Lara, Whiteford, and García Acevedo 2000) on actual events surrounding the lining the of the AAC indicate there is no formal evidence of proposed solutions that have been fully discussed among the stakeholders. Only in a few cases has this discussion taken place at the level of the authorities of both countries related to this issue, namely, IBWC and its Mexican counterpart, Comisión Internacional de Límities y Aguas (CILA).

Since 1988 the subject has been raised intermittently by U.S. authorities and the Mexican side has reacted, albeit without any cohesion that would support a uniform proposal and that could be seen as a negotiation position that could bring a balanced solution to the conflict. In this sense, for example, at the federal level, CILA does not have consistent studies such as those IBWC has and uses for binational negotiation and consultation with Mexico (for example, the Final Environmental Impact Statement and the Final Environmental Impact Report for the AAC lining project). This speaks to the need to strengthen technical aspects on the Mexican side so they can lay a solid basis for negotiation and put leaders in a position to give solid, sustainable answers to IBWC.[10]

With Mexican society perceiving the response from their representative authorities as lukewarm, isolated, and generally unsatisfactory, undoubtedly different stakeholders will voice their concerns, though probably in the same manner as the government has. Such was the case with the recent technical recommendation (Cortez, et al. 2002) that the Baja California state government requested in order to develop its position before authorities in both Mexico and the United States.[11]

Other government stakeholders have also tried to get involved with the canal's lining. Starting in 1998, the various administrations of the municipal government in Mexicali, together with Imperial County, discussed compensation alternatives—referring to the possibility of having water from the Colorado River delivered through the AAC under the following guidelines:

- A portion of the water allotted under the 1944 Water Treaty (Lindero Norte/Morelos Dam) would be delivered at a point near the Mexicali urban area
- Mexico would build the structure to convey the water from the point of delivery to the city of Mexicali
- A volume of up to 500,200 AF/y would be conveyed

- Mexico would not be charged for conveyance losses along the 31-mile route
- The quality of the water delivered to Mexico would be similar to that of the water conveyed through the AAC
- Mexico would cover all costs incurred

This proposal was deemed infeasible by CNA for technical contradictions. Among them were the fact that water salinity concentrations at the original point of delivery set forth in the treaty would result in negative consequences for the rural population and agriculture in northeastern Mexicali Valley.

Other alternatives in addition to this one have been proposed, including modifying the original project,[12] using the Yuma desalination plant, and delivering water from the Colorado into Tijuana. Still other alternatives may be proposed, but stakeholders must attempt to establish a conflict-management strategy based on equilibrium of the solutions and, as a result, of shared benefits.

The example of the alternative of delivering water to the city of Mexicali through the AAC makes it clear that the efforts of one of the stakeholders (the municipal government, in this case) to attract a benefit for the city of Mexicali did not succeed, or at least has not yet, because it did not take into account the negative effects on third parties (in this case, the rural stakeholders or agricultural producers). It once again fell prey to playing with the conflict without including all the stakeholders involved. This, in the end, makes it more difficult to reach a weighted solution that benefits both countries.

Conclusions

While the condition of the aquifer has been determined by CNA, Mexico lacks the necessary technical and environmental impact studies that would allow for a fair negotiation with its counterparts. Information exists that creates enough suspicion to respond negatively to the project, but a formalized answer is extremely advisable. The pending environmental impact studies do not necessarily have to be conducted by government agencies, but what they would need is a great deal of public participation. This is where integration should be sought, first as a Mexican region, including social stakeholders such as Sociedad del Distrito de Riego (Irrigation District Society), which represents the nearly 14,000 agricultural users; the five irrigation districts or user associations that could be directly and immediately affected (districts 4, 5, 6, 7, and 16); the residents of surrounding communities; the business sector; environmental groups; academia; and the corresponding agencies from the different levels of government.

On another front, the challenge is to bring U.S. and Mexican stakeholders together, especially to exchange information, homogenize criteria, and discuss both technical and socioeconomic issues that unquestionably affect both sides because of their degree of regional interdependence. The goal of benefiting more than one party should be promoted in both nations. By incorporating an equilibrium approach like the one proposed by Nash, the potential for reaching a solution with a positive net value that includes every stakeholder and that, as a result, minimizes affectation of third-parties, would be increased.

A short- or medium-term solution resulting from the implementation of a cooperative model could include actions aimed at making the binational regulatory framework for the management of cross-border groundwater more suitable. Given the existing interdependence and competition for a resource with a high exclusion cost, Nash's negotiator/conciliator model can address the conflict over the AAC, if suggestions made by Schmid are included to develop an intensive and iterative dialogue negotiation process.

Cases such as the AAC could occur more frequently along the U.S.-Mexican border given the demographic and economic dynamics that exist in the region, as well as the growing water demand. Therefore, and with the understanding that development cannot be stopped and that the economic growth of one region generally has positive effects on the other, it is essential to adjust the rules of the game for crossborder groundwater management so that an efficient, equitable, and sustainable use of the resource may be achieved.

Endnotes

[1] The Colorado River is divided into the upper and lower basins at Lee's Ferry, which is located a few miles south of Lake Powell and its respective Glenn Canyon Dam. This dam supplies, downriver, parts of Nevada, Arizona, New Mexico, and California, whose normal allocation is 7.5 MAF/y, as set forth in the Colorado River Compact approved by the U.S. government in the early 1900s.

[2] Population estimated based on 4.2% average population growth projected for 2000 to 2010 (IID 2001).

[3] Note that on a watershed level, 85% of the water volume used is still allocated to agriculture. In more arid regions, such as California, this percentage is 92%.

[4] By the reduction of underground volumetric availability alone, caused by keeping the waters of the AAC from seeping into the Mexicali Valley aquifer, the surface area for agriculture would be reduced by 9,149 ha. However, because of indirect effects due to the volume-quality ratio of these waters, the damage could potentially expand gradually to 33,400 ha.

[5] Data to June 30, 2001, considering a 2.3% average annual growth rate observed from 1995 to 2000 (Consejo Estatal de Población 2004).

[6] The environmental impact assessments made in these years started out with the initial position, in 1988, of not disclosing results about potential crossborder impacts for Mexico. In 1994 it was done more openly but without including opinions from the Mexican section. In the recent National Environmental Protection Act process in 2002, which produced the new Environmental Impact Review and Environmental Impact Statement for the AAC lining project, it was opened to public comment in Mexico. Yet, there was no opposition to the project, other than CILA's official position expressing disagreement. However, this statement was made without any technical documents to support it.

[7] The most recent and strong reaction opposing to the AAC initiative is the legal action against the U.S. government by a few local non-governmental organizations established on both sides of the border, including Consejo de Desarrollo Económico de Mexicali

(CDEM), United Citizens for the Natural Resources and the Environment (CURE), and Citizens of the Desert Against Pollution (CDAP).

[8] The idea presented in this chapter regarding the model universally spread by classical economists such as Adam Smith, neo-classical economists like Alfred Marshall, and new institutionalists like Oliver Williamson, is that the government should play only a regulatory role and impose as few restrictions as possible on trade and pricing.

[9] The four mathematical principles behind the modern "Game Theory" are developed at length in the four documents published by its author, John Forbes Nash. In his essays, Nash makes basic, valuable contributions both to Uncooperative Game Theory and to Fair Bargaining Theory (*Essays on Game Theory*, John Forbes Nash, 1997).

[10] According to Official Document LAE 01012/00, drafted in Ciudad Juárez, Chihuahua, on May 22, 2000, the Mexican commissioner extends a Diplomatic Note to his American counterpart expressing his disagreement with the AAC lining project, citing the potential negative effects it could have on the Mexicali Valley.

[11] This recommendation, based on a series of technical documents prepared by agencies from both countries, essentially establishes that the Mexican government should oppose lining the AAC until they have:

- A complete study on the socioeconomic and environmental impacts of the canal lining on the Mexicali Valley
- Alternatives to mitigate the effects lining the canal would have in Mexico

[12] It is important to stress that this alternative is more of a recommendation borne of the environmental study conducted in the United States, which was then translated into Spanish and presented as a negotiable option for Mexico. In fact, the project was modified due to pressures from U.S. environmental non-governmental organizations.

REFERENCES

Comisión Nacional del Agua and Gerencia Estatal en Baja California. 1991. *Efectos del Revestimiento del Canal Todo Americano sobre Territorio Mexicano*. Mexicali, B.C.: Comisión Nacional del Agua and Gerencia Estatal en Baja California.

Comisión Nacional del Agua and Gerencia Estatal en Baja California. 1999. *Recomendación técnica de la propuesta de entrega de agua del CTA a la ciudad de Mexicali*, B.C. Mexicali, B.C.: Comisión Nacional del Agua, Gerencia Estatal en Baja California.

Consejo Estatal de Población. 2004. *Proyecciones de población por localidad*. Mexicali, B.C.: Gobierno del Estado de Baja California.

Cortez Lara, A. A. 1999. "Dinámicas y conflicto por las aguas transfronterizas del Río Colorado: El proyecto All-American Canal y la sociedad hidráulica del Valle de Mexicali." *Frontera Norte* 11(21): 33–60.

Cortez Lara, A., and S. Whiteford. 1996. "Conflictos urbano/rurales sobre el agua del Río Colorado en el ámbito internacional." Pages 121–153 in *Agua: desafíos y oportunidades para el siglo XXI, Memorias del Seminario internacional sobre Gestión Eficiente del Agua*. Aguascalientes: Gobierno del Estado de Aguascalientes.

Cortez Lara, A., S. Whiteford, and M. R. García A. 2000. "The Lining of the All-American Canal: The Forgotten Voices." *Natural Resource Journal* 40(2): 261–279.

Cortez Lara, A., A. López L., M. L. López, J. Mosqueda M., E. Paredes A., and A. Sandoval S. 2002. *Recomendación técnica respecto al revestimiento del Canal Todo Americano*. Mexicali, B.C.: Gobierno del Estado de Baja California.

Cropsey, J. 1977. *Policy and Economy: An Interpretation of the Principles of Adam Smith*. Chicago: University of Chicago, Greenwood Press.

Frank, R. H. 1988. *Passions Within Reason: The Strategic Role of Emotions*. New York: W.W. Norton.

Frisvold, G. B., and M. F. Caswell. 2000. "Transboundary Water Management: Game-theoretic Lessons for Projects on the U.S.-Mexico Border." *Agricultural Economics* 24: 101–111.

Howe, Charles. 1997. "Increasing Efficiency in Water Markets: Examples from the Western United States." Pages 79–99 in *Water Marketing, The Next Generation*, Terry L. Anderson and Peter J. Hills, eds. Lanham, Md.: Rowman & Littlefield.

Imperial Irrigation District. 2001. *Profile of the Imperial Irrigation District 2001*. Imperial, Calif.: IID.

Ingram, H. 1992. *The US-Mexico Border Region under Stress. A Binational Symposium on Ideas for Future Research. Final Report, Proceedings and Recommendations*. Guaymas, Son.: Ford Foundation, Udall Center for Studies in Public Policy and University of Arizona.

International Boundary and Water Commission. 1973. *Minute 242 Permanent and definitive solution to the international problem of the salinity of the Colorado River*. http://www.ibwc.state.gov/Files/Minutes/Min242.pdf.

Kuhn, H. W. 1996. *A Celebration of John F. Nash*. Boston: Princeton University Press.

Navarro, U. J. 1998. *Impacto del revestimiento del Canal Todo Americano en el Distrito de Riego 014, Río Colorado, Baja California y Sonora*. Mexicali, B.C.: Universidad Autónoma de Baja California.

Platt, J. 1973. "Social Traps." *American Psychologist* August: 641–650.

Schmid, A. A., 2004. *Conflict and Cooperation. Institutional and Behavioral Economics*. Malden, Mass: Blackwell Publishing Ltd.

Trava M., J. L., J. R. Calleros, and F. A. Bernal R., eds. 1991. *Manejo ambientalmente adecuado del agua. La Frontera México-Estados Unidos*. Tijuana, B.C.: El Colegio de la Frontera Norte.

U.S. Congress. 1998. *Public Law* 100-675, The Lining of the All-American Canal.

U.S. Department of Interior, Bureau of Reclamation, Imperial Irrigation District. 2002. *Final Environmental Impact Statement/Final Environmental Impact Report*. All-American Canal Lining Project. Calexico: Bureau of Reclamation.

X

Context and Implications for Resolving a Complex Binational Issue: Lining the All-American Canal

Vicente Sánchez Munguía

Introduction

The U.S. government's 1998 decision authorizing the All-American Canal's (AAC) concrete lining has become one of the most complex issues in international water negotiations between the United States and Mexico in the last several years—the fact that it has not been given as much attention by government entities, academia, and public opinion as other issues notwithstanding. This is not the first time both countries face the challenge of finding common ground for resolving their differences on water-related issues. However, the problem that lining the AAC represents, and its consequences for Mexico, are linked to strong interests by a series of public and private parties that seek to ensure access to a scarce resource that has no substitute and faces growing demand. It is this last issue that has to a large extent determined the conditions under which negotiations between the two parties have taken place.

These negotiations are being held in an atmosphere marked by tremendous regional competition and involving several users and entities that seek to ensure their future viability via ensuring access to, and control of, the region's scarce water resources. Moreover, these resources are subject to formal delivery agreements among the various jurisdictions. That is, these waters are carefully allocated among each of the two countries' border states—the pressure exerted on the resource manifests as an intense competition among the different entities involved in its management, all of whom are trying to obtain as large an allocation as possible.

Although the lining has been delayed for various reasons, the decision to undertake the project in the first place has not undergone any fundamental change. Meanwhile, negotiations seem to have reached an impasse as a result of the opposing definitions of the problem by each of the parties. However, as this chapter will demonstrate, the Mexican section of the International Boundary and Water Commission (IBWC), Comisión Internacional Límites y Aguas (CILA), has given some indication of a potential agreement on the lining and the water losses for the Mexicali aquifer.

Both countries have held fast to their different definitions of the problem, which is understandable, and have held what would seem like opposing postures that would be difficult to reconcile. From the California and U.S. perspectives, these waters belong to them under the allocations set forth in the Treaty on the Utilization of the Colorado and Tijuana Rivers and of the Río Grande between Mexico and the United States, signed in 1944 and referred to as the 1944 Water Treaty. A portion of this water is lost to seepage as it is conveyed through the AAC. In keeping with this definition, the United States can sovereignly decide on the management and allocation of this resource. As a result, U.S. officials have assumed they can move forward with lining the AAC as a method to recover their lost water, and then export it to the Greater Los Angeles Area.

Meanwhile, Mexico's position on this same issue has been based on the fact that stopping the groundwater flow into the Mexicali Valley would violate Mexico's rights (Bustamante 1999). That is, these are waters over which Mexico has created a right by continuous use over a long period of time.[1] As well, Mexico has pointed to the violation of point 6 of Minute 242, which establishes that:

> With the objective of avoiding future problems, The United States and Mexico shall consult with each other prior to undertaking any new development of either the surface or the groundwater resources, or undertaking sub-

> stantial modifications of present developments, in its own territory in the border area that might adversely affect the other country.

While it seems Mexico's claim centers on the lack of consultation as set forth in this point, note that its first line establishes the goal of the agreement—avoiding future problems. However, the most relevant aspect, set forth at the end of the paragraph, provides that potential damage to the other country from a particular action should lead to actions to avoid such negative impacts in the future.

Nevertheless, the clarity with which this agreement should have been understood at its signing in 1973 has not been exhibited in putting it into practice. At the time, the aim was to definitively resolve the problem of the quality of water received by Mexico, which had caused serious damage to Mexicali Valley agriculture. Today, Mexico seeks to avoid the damage that lining the AAC would cause the Mexicali Valley from the water loss that would result from it.

So, what was once agreed upon is now interpretable. Not only is the concept of consultation (lack of notification) subject to interpretation—aligned with the interests being defended by each side—but also the substance of Minute 242's point. If these issues are forgotten or cast aside, the agreement set forth in this point of Minute 242 would lose its essence and be left with no force or effect.

However, beyond any legal or doctrinal considerations, or any interpretation made on the point in question, lining the AAC would certainly have a direct effect on the recharge of the aquifer from which agriculturalists in Mexicali Valley's Irrigation District 014 extract their water, subsequently endangering their livelihoods and families. The lining would presumably cause significant damage to the irrigation surface area,[2] where the water now in dispute has thus far been used.

At the heart of the differences of opinion lies the fact that any solution proposed by the parties must be satisfactory for both and avoid a negative impact on Mexicali Valley growers. In this sense, this chapter analyzes the problem and proposes that the only likely means to overcome the impasse in negotiations is by integrating other elements of the water-management problem in the region. The current negotiation scheme offers few possibilities, given the asymmetry that characterizes the relationship between the two countries (Sánchez 1990). However, recent history shows significant progress in binational environmental cooperation, and it would be in this arena that the issue of lining the AAC could find a mutually agreeable resolution.

This chapter has three parts. The first is a general overview of the problem, based on a contextualization of the relations established among different public and private entities created for accessing and controlling water in the region and beyond. The second section focuses on the perception in certain Mexican political spheres of the lining of the AAC. Finally, some of the possible means of resolving the conflict are discussed and some overall conclusions offered.

The Border and the Waters of the Lower Colorado River Basin

The conflict between the United States and Mexico over the AAC lining is nothing more than a reflection of the tension created by the intense competition among water users in the Lower Colorado River Basin, particularly between California and Arizona. As well, it brings to light the water quality problems of the Mexicali-San Luis Río Colorado area.

Under the Colorado River Compact, California's allocation from the Colorado River is 4.4 million acre-feet per year (MAF/y). However, because Arizona and Nevada did not use their allocations for many years, California used their water, thus exceeding its own allocation by at least 800,000 AF/y. But in 1968, after 20 years of intense lobbying before Congress and the U.S. Department of the Interior, U.S. President Lyndon Johnson authorized the Central Arizona Project (CAP)[3].

Since the mid-1900s, CAP had been a strategic goal for the most politically influential Arizona groups. Designed to convey 1.5 MAF/y of Colorado River water to central and southeastern Arizona, CAP was the product of a long struggle by Arizona to secure its access to Colorado River water and was for a long time a source of dispute with California. Even back then the competition between these states for Colorado River water was keen. But, although the original CAP project was meant to convey water for agricultural use in Central and Southeastern Arizona, the urban demand imposed by the growth of cities such as Phoenix led to modifications[4] to meet the needs and demands of an urban dynamic that had not originally been considered. Thus, the water from CAP became primarily an urban-use resource. It now counts 80 large users among its customers, 75% of which are municipal and industrial users, 13% are irrigation districts, and 12% are tribal nations (Central Arizona Project 1997). Municipal and industrial customers are allocated just more than one-third of the water from CAP (555,031 AF/y) (Central Arizona Project 1997).

Construction of CAP finished in 1993 and with it began the diversion of water allocated to Arizona. California counts this diversion as a loss of water it had been using for a long period of time, water that had met the growing demand in Southern California cities, particularly Los Angeles and San Diego. During this same timeframe, the federal government mandated that California reduce its dependence on Colorado River water so that by 2002 it could return to its allocated level of 4.4 MAF/y. These decisions were seen by California as a threat to its future water supply.

California is feverishly seeking a future water source, especially to serve Los Angeles and San Diego. Northern California has a greater abundance of water, much more than the eastern portion of the state. However, voters rejected a proposal to transfer water to Los Angeles from the San Joaquin Valley and the Sacramento River in a 1982 referendum. The fact is, both CAP and the rejection of water transfers from Northern California have created a state of anxiety among those drafting regional water policies.

Framed by this trend, water management agencies in urban areas have begun searching for alternative future supplies. The San Diego County Water Authority and the Imperial Irrigation District signed an agreement in 1998 for the transfer of 200,000 AF/y for a 45-year period, with the possibility of expanding the agreement for an additional 30 years at the end of the 45-year term (SDCWA No Date), effectively ensuring San Diego's water supply for the remainder of the 21st century.

It was this same spirit that developed the proposals to line the All-American and Coachella Canals (the Coachella is also in Imperial Valley). All these water saving and transfer projects aim to replace some of the water lost by California to CAP (Hayes 1991) and are linked to Plan 4.4, which was created to comply with the mandate to reduce dependence on Colorado River water. The Metropolitan Water District and the Imperial Irrigation District agreed to line a 30-mile stretch of the AAC and a 38-mile stretch of the Coachella Canal, contributing part of the overall cost so they could contract with the federal government for 55 years for the purchase of the rights to the water saved (Hayes 1991).

This water savings and transfer policy, however, is highlighting the competition that has historically existed between California and its neighbors, Arizona and Mexico, for Colorado River water. The competition now involves Phoenix and Los Angeles, metropolitan areas with large water demand, and pits urban users against agricultural users.

Meanwhile, the region's water scarcity in the face of a growing urban demand gives this competition peculiarities worth noting, given the environmental impact they have on the lower portion of the Colorado River Basin. For example, urban demand in Southern California and Baja California comes mostly from areas outside the watershed's limits (except for Mexicali), which means these cities located at considerable distances from the bed of the river become significant water importers. This translates into a net loss of water for the watershed, with its corresponding environmental impact, because the river receives no return flows after the water is used in these far away places.

Although the demographic dynamic is not an entirely new phenomenon, and neither is the growing demand for water imported from the Colorado River, the fact is the growth in demand in a region characterized by the scarcity of the resource can become, and is in fact becoming, a source of binational tension and conflict. The resolution will require determination and imagination. The growth dynamic reached by the urban systems demanding water from the Colorado is a relatively new phenomenon,[5] and it is difficult to fathom San Diego, Tijuana, and Ensenada having the same dynamics without the water they receive from the Colorado, on which they depend to a significant degree.

Urban development in the Southern California (mostly San Diego) and Baja California coastal region has been possible only with the import of large volumes of water from the Colorado River. The region's growing water demand on both sides of the border is actually outside the limits of the Colorado watershed, which means the imports cause a net loss of the resource and a significant environmental impact for the ecosystem from which the water is taken.[6]

Taking this trend into account, and given that the water in dispute is extracted from binational watersheds, substantial modifications have been suggested to the current water management schemes in the border region. The goal is for these management schemes to evolve into binational cooperation models that will overcome potential conflicts and benefit the residents of both countries (Nitze 2002).[7] The basis for such a model is Elinor Ostrom's (2000) proposal regarding "the commons," in which cooperation is the center of the process for using shared resources to avoid competition over appropriations and ensure its future viability through their protection (Ostrom 2000).

Obviously, an effort like this requires a different institutional framework than what has prevailed for the binational management of resources whose sources are traversed by political boundaries that limit the jurisdiction of laws and governmental action. The state-centric vision that has characterized border management by the United States and Mexico is not keeping pace with changes in cross-border relations, where the intensity of exchanges and interactions on both sides of the border have woven a network of multiple interinstitutional contacts and exchanges relating to water policy (Blatter and Ingram 2000).

In this sense, the last decade has borne witness to notable changes toward environmental cooperation, with laudable participation from the agencies created in parallel to the North American Free Trade Agreement (NAFTA) and a substantial increase in crossborder contacts and exchange, both on a social level and between government entities. In spite of this progress, the sovereignty approach on which the clamors over water appropriation and management are supported persist on both sides of the border, as if nothing has changed. However, it is important to note that the recurring and prolonged droughts in the region, as well as the growing water demand, have caused concern in the United States and Mexico, leading both governments to turn their gazes toward reaching greater cooperation as a means of achieving efficient water management objectives. This would require better coordination, a common policy on watersheds and surface and groundwater, and joint planning for medium and long term use of resources, as well as investments and mutual assistance for resource conservation.

Mexico'sOfficial Position on Lining the AAC

Although the U.S. government's decision to line the AAC has been communicated to Mexico since 1976, the Mexican government's response has been cautious—or low-profile—in spite of its opposition to the project's implementation. Discussions on the issue have been limited to academic circles. The Mexican Senate, though, because of its dealings in matters of foreign policy, has more eloquently voiced its opposition. However, the lack of information passed from the Mexican government to other public agencies and the public at large has limited the ability of other entities to publicly take a position on the matter. This is what Baja California Senator Norberto Corella was referring to when he spoke of public officials in charge of executive-branch agencies hiding information from the Senate.

Limited information on this issue in Mexico has been disseminated because of the interest (perhaps somewhat belatedly) of northern Mexican border state senators. It has been the legislators of Partido Accion Nacional's parliamentary group who have more forcefully addressed this issue. On April 11, 2000, Baja California Representative Fortunato Álvarez Enriquez proposed before the Mexican Chamber of Deputies Rostrum a resolution asking that the chamber's Border Affairs Commission call the officials in charge of Secretaría de Medio Ambiente y Recursos Naturales (SEMARNAT), as well as those from Comisión Nacional del Agua (CNA, in English

National Water Commission) and CILA to appear and inform the commission about the AAC lining project and ways to avoid the damage it could cause Mexico.[9]

In response, the Senate called a meeting with officials from the Secretariat of Foreign Relations, CILA, and CNA, "...to become acquainted with the status of measures being undertaken by Mexico to mitigate the effects of the lining of the All-American Canal" (Mexican Senate 2000). The Secretariat of Foreign Relations stated that the legal basis on which the Mexican government has based its opposition is the 1944 Water Treaty and Minute 242, which establishes the right to mutual consultation in decisions that affect the other country (Mexican Senate 2000).

Examining the opinions expressed during that meeting, it can be concluded that the Mexican government does not have an inflexible position or a desire to reject the project outright—indeed, approaches like these could exacerbate the conflict. This is evidenced by the participation of several senators, and in particular Corella, who pointed out that there is a legal basis for the Mexican government to intervene and impede the lining of the AAC, but added that a favorable agreement could be reached, even though the entire project is to be carried out within U.S. territory.[10]

In this same vein was the intervention by CILA commissioner Arturo Herrera Solís, who told senators that as early as 1991 Mexico had expressed its opposition to the project because of the negative effects it would have on its territory. He committed to legislators that he would identify joint measures to mitigate impacts, as well as request that mutually beneficial alternatives be sought. One of the options Herrera raised for mitigating the impact of the lining would be for Mexico to receive its water apportionment through the AAC and explore measures to preserve and protect the aquifer, as well as legislate the mandatory recharge of all water surpluses to the aquifer.[11]

Additionally, the Border Affairs Commission organized a meeting in Mexicali in May 2000 of a broad spectrum of stakeholders including representatives from the main agencies involved, local authorities, water users, and academics from the region.[12] Although there was a tone critical of the U.S. unilateralism and of the interim agreement granting California surplus water from the Colorado

River, a substantive statement came from the CILA Commissioner, who said the state of California planned to build a new AAC to save more water and to use the old canal to convey water for environmental uses in the U.S. border region. He presented some the elements of the proposal that CILA requested, including:

- That the new canal be used to convey water for delivery to Mexico, which after going through the sediment-control plant would allow Mexico to receive higher-quality water and would eliminate sediment-control costs for urban uses
- That part of the surplus California intends to save be conveyed through the old canal so that water can be allocated to Mexico and seepage be allowed that Mexico can store for later reuse
- Mexico wants to use its water conveyed through the AAC to generate electricity, and the economic benefit of this to be designated as payment for the infrastructure needed by farmers who are affected by the development of the new AAC[13]

The foregoing notwithstanding, according to Baja California Governor Eugenio Elorduy Walther, the Mexican federal government wants the AAC to remain unlined and, by means of a diplomatic memorandum to the U.S. Department of State by the Mexican embassy in Washington, wants to request formal discussions on the topic.[14]

It would appear that the arrival of the Fox Administration and the design of a new foreign policy agenda have brought an about-face in how the AAC problem is handled. CILA had seen the lining decision as irreversible and had thus focused on finding potential alternatives for the mitigation of impacts on the Mexican side of the border. However, it never reached the point of discussing a specific proposal to be officially presented to its U.S. counterpart, and thus what the CILA Commissioner had expressed to the senators seemed to be somewhat speculative. But with the Fox Administration now in full swing, CILA has recognized, as expressed by its own officials, that this is an issue that is no longer in their hands and that will be addressed through diplomacy (Pensamiento 2002). This means the Secretariat of Foreign Affairs will negotiate directly with the U.S. government.

If this change is true, and negotiations have indeed been elevated to the federal level, it would mean the Mexican government is convinced that the negotiation route previously followed was not the most efficient one. It would be the responsibility of this secretariat to convince the U.S. government not to line the AAC, something that seems extremely difficult, considering the U.S. Bureau of Reclamation (BOR) announced construction would begin in 2003 in light of an investment of $200 million (Pensamiento 2002).

Although the execution of the project has been delayed time and again, there has been no indication that the decision will be rescinded definitively[15] or that the project will be modified. On the contrary, BOR officials have made it clear that the decision is final, given that these are U.S. waters and are not subject to the 1944 Water Treaty. However, they have acknowledged that the execution of the project will damage wells on the Mexican side of the border and in that regard have expressed a willingness to discuss claims.[16]

The foregoing demonstrates that negotiations are at an impasse and that the problem is back to square one, where positions are rigid and there is little willingness by the parties to change their points of view and reach an agreement. Keep in mind that water issues have gained ground on the bilateral agenda, but in recent years have become more complex in the relationship between the two countries, due to Mexico's delay in paying back water from the Rio Grande to the United States. The fact that the Fox Administration's Secretariat of Foreign Relations has taken up negotiations directly is indicative of a repositioning of the Mexican government's definition of the problem.[17]

Lining the AAC and Other Measures

As discussed in the preceding section, CILA had assumed lining the AAC was a *fait accompli*, and thus proposed potential alternatives to mitigate or avoid impacts on the Mexicali Valley aquifer. It is not clear whether any of these measures had been discussed with IBWC, and therefore, nothing is known about any progress on the acceptance of these proposals by the United States. Some of the proposals seem logical and would not entail great sacrifices by the parties, particularly if the current canal will convey water for environmental uses. This means it would eventually convey water allocated to Mexico at a newly agreed upon point of delivery, and as a result part of the water would seep into the aquifer and avoid damage to growers. Another possibility would be for Mexico to receive higher-quality water as a compensatory measure for the damages lining the canal would cause the aquifer. Nothing has been said about measures to mitigate impacts on the farmers who use this water.

At the same time, to date the parties have analyzed only the direct benefits and damage lining the AAC would cause. They have not examined the problem within the context of overall problems in the region stemming from overexploitation of the Colorado River—the project to rescue the Colorado Delta and the one to protect the Salton Sea are not seen as interconnected with the lining of the AAC.[18] These two projects are essential for maintaining ecological balance in the Lower Colorado River Basin and protecting biodiversity in these ecosystems. Fresh water is absolutely necessary to preserve the minimum conditions for survival (Asociación Ecológica de Usarios del Río Hardy-Colorado, A.C., et al. 2001).

This partial, unilateral vision by California and the Los Angeles and San Diego water utilities has remained unchanged throughout the process. On September 25, 2003, the San Diego County Water Authority and Imperial Valley agricultural users reached an agreement to transfer 200,000 AF/y for 75 years, at the same time SDCWA accepted the 77,000 AF/y for 110 years offered by the Metropolitan Water District. This water would come from savings after lining the AAC.[19] Concurrently, then-California Governor Gray Davis signed laws supporting the transferring of water from Imperial Valley to San Diego. Although these legal provisions allocate resources for restoring the Salton Sea, nowhere do they mention the Colorado River Delta as an affected user, nor do they make reference to the impact on Mexicali of lining the AAC, whose cost will be covered by the San Diego County Water Authority in return for the transfer of the saved water.

Additionally, as part of the border remediation program supported by binational cooperation proponents, significant investments have been made to clean up the New River (Río Nuevo) in Mexicali and to treat the wastewater it conveys to avoid environmental and public health problems on both sides of the border.

Eventually, Mexico could reclaim this water and use it differently, such as for environmental uses and aquifer recharge, and thus maintain static levels in the wells while improving water quality. However, a reduction in its flow in U.S. territory could have a negative impact on the Salton Sea, so specific provisions would be necessary if the problem is to be addressed integrally and jointly, reaffirming the idea of fully adopting a water management model in the region in accordance with the interdependence of the ecosystems currently fragmented by the intense intervention of both societies.[20] Several non-governmental organizations and research institutions on both sides of the border have made progress on proposals to build a framework for the joint, integral management of the ecosystems of which the water in dispute is a part. Thus, it is the political arena that lacks approaches to cooperation and mutual concession that could result in better management, from the environmental perspective, as well as in greater social and institutional stewardship.

Conclusion

The pressure being exerted by water demand on the Lower Colorado River Basin is a factor that exacerbates the binational differences and conflicts surrounding this resource. Over the years, the United States and Mexico have found ways to resolve conflicts. However, the margins within which each government operates are ever narrower, given that the objective of those involved in the complex bidding for water is to obtain the largest volume possible, and that the game takes place under conditions where all the water is already allocated. As a result, pressure is exerted upon those areas and sectors of consumers who have relatively larger water allocations or volumes (such as the agricultural sector), with the aim of introducing water savings and transfer policies. In the case of the border region, this could lead to potential collateral effects and implications on crossborder water relations.

Additionally, the lack of a binational relationship that transcends the unilateralist approaches anchored in national sovereignty postures, as in a state-centric model, have led to fragmented, contradictory policies even within the countries themselves. Thus, there is a disconnect between the AAC-lining project, the Salton Sea

Preservation project, and the Colorado Delta project, all of which are located in the same area and should therefore be approached with a policy that provides more congruency to the goals of each of the individual projects. To this add the signing of an interim agreement that apportions Colorado River surplus waters to California for a 15-year period beginning in 2002.

The arrival of a new government in Mexico in 2002, with a more prominent foreign policy agenda and with perspectives of leaning on the "democracy bonus" to negotiate with the U.S. government, led to the Secretariat of Foreign Relations directly assuming control of the AAC lining issue, removing it from CILA's purview. However, several factors, including the U.S. protesting Mexico's historical water debt with Texas, have shown that democratic legitimacy was not enough to ensure a successful negotiation with the United States. At the same time, the Mexican Senate has disapproved of the method the Mexican government used to address the binational subject of water. The legislators now express doubt and reservations as to how the Fox Administration has handled these types of problems and accuse the president of favoring U.S. interests. This has made finding a solution to an already difficult situation even more complicated.

Seeing the complexities of the AAC lining problem and the difficulty in arriving at a solution, the question is, Does an avenue for resolving the conflict even exist? Surely one must, but it is more likely found in incorporating all elements of the ecosystem from which the water is taken and adopting a different approach to cross-border water management. In this regard, society on both sides of the border has made greater strides than governments. Finally, the agreements reached by stakeholders within California (namely, the water utilities) and the laws signed by the California governor to support water transfers from the Imperial Valley to San Diego (including lining the AAC) indicate that resuming negotiations has become more complicated for Mexico, given that actions to that end have already been taken by California, with the support of the U.S. Department of the Interior.

ENDNOTES

[1] On this matter, a thesis by Albert Utton is frequently quoted that states international precedent would support the right acquired by Mexico as the user of the disputed water during more than 25 continuous years, thus leading to the loss by proscription of the Imperial Valley Water District's rights. But he also recognizes how difficult it would be to reach a resolution in court, due to a lack of international treaties and agreements regulating the use of groundwater. See Bustamante (1999) and Utton (1991).

[2] There is no agreement as to the surface area that would be affected by lining the AAC, indicating that more studies and data are needed to more precisely assess the magnitude of the impact on the agricultural sector in the area. Hayes (1991) speaks of 13,500 hectares, as does Calleros (1991), Navarro Urbina, however, quoted by Cortez Lara and García Acevedo (2000), speaks of 19,200 hectares. CILA Commissioner Arturo Herrera Solís refers to 1,200 hectares (Mexican Senate 2000). It is 9,200 hectares according to Senator Norberto Corella, who quotes Operations Management for Irrigation District 014 (www.pansenado.org.mx/Debate/cona000406a.html). Member of the Chamber of Deputies Fortunato Alvarez Enríquez (www.gaceta.cddhcu.gob.mx/Gaceta/2000/abr/20000412.html) uses this same figure. Mendoza (2002) states that the surface area irrigated with water extracted from the ground in that part of the irrigation district is 70,000 hectares.

[3] "Central Arizona Project is designed to bring about 1.5 million acre-feet of Colorado River water per year to Pima, Pinal and Maricopa counties... It is a 336-mile long system of aqueducts, tunnels, pumping plants and pipelines and is the largest single resource of renewable water supplies in the state of Arizona" (Central Arizona Project 1997).

[4] In 1977, the Arizona Water Commission, in a letter to the U.S. Secretary of the Interior, announced recommendations for CAP water allocation to municipal and industrial users (Central Arizona Project 1997)

[5] From 1940 to 1990, Mexicali's population increased 2,209%, Calexico's grew 198%, Tijuana's grew 4,138%, and San Diego's grew 764% (Castro 2002).

[6] Although there are legal provisions in both countries restricting water extraction to the point where it competes with and endangers species and ecosystems, thus far neither has been very strict in overseeing and enforcing their respective legal precepts. (Nitze 2002).

[7] Several papers from academics have suggested the possibility of reaching cooperative methods and binational agreements to achieve proper, sustainable water management or for restoring the degradations of the ecosystems caused by the overexploitation to which they have been subjected. The proposals have been generally based on the idea of "the commons." See Pitt, et al. (2000), Brown and Mumme (2000), Michel (2000), and Nitze (2002).

[8] In fact, it was Corella who brought the issue before the Senate Rostrum during an intervention on May 6, 2000, before the purportedly imminent beginning of construction to line the AAC. During his intervention, the Senator proposed a resolution asking the Secretariat of Foreign Relations to send a diplomatic memorandum to the U.S. Embassy requesting that the lining of the AAC not begin as scheduled, and second, to ask Secretaría de Medio Ambiente y Recursos Naturales (SEMARNAT) to appear before the senators and say whether it had been consulted about lining the AAC, per the provisions set forth in the corresponding international treaties. www.pansenado.org.mx/Debate/cona000406a.html

[9] www.gaceta.cddhcu.gob.mx/Gaceta/2000/abr/20000412.html.

[10] www.pansenado.org.mx/Boletines/bole801.html.

[11] Ibid.

[12] www.senado.gob.mx/comunicación/boletines/2001/b26mayo.html.

[13] Ibid.

[14] In that same regard, an article by Angélica Enciso in a Mexico City newspaper states that in 2001, the Mexican Office of Foreign Affairs sent a diplomatic memorandum to the U.S. government, but that the evaluation period is over and lining the AAC is now a reality. In the same article, a regional representative of CILA is quoted as saying that the issue is in the hands of the Secretariat of Foreign Affairs. The article appeared in the April 15, 2002, edition of *La Jornada*.

[15] The lining was scheduled to begin in 2003, but was postponed again, according to statements by Engineer Leonel Vizcarra Ojeda, Director General of the Baja California State Water Commission, before the State Congress in October 2002.

[16] This is according to an article by Angélica Enciso in the April 15, 2002, edition of *La Jornada*.

[17] It is worthwhile to note that the pressure exerted by the U.S. government on Mexico for payment of its water debt, and the activism adopted by the Mexican government in its foreign policy, have created notable differences within the Mexican Senate, including differences in interpretation of the 1944 Water Treaty, which in some ways could be reflected in Mexico's position before the United States in negotiations for resolving water issues, including lining the AAC.

[18] The Pacific Institute Report on the Salton Sea establishes that it is part of a larger regional ecosystem and therefore its restoration must be compatible with broader, long-term efforts to restore the Colorado River Delta and the Upper Gulf of California eco-region. (Cohen, et al. 1999).

[19] This is according to an article in the September 26, 2003, edition of the *San Diego Union-Tribune*.

[20] Cohen, et al. (1999) have stressed the critical relationship between water flows in the region and AAC lining, New River remediation, and Salton Sea restoration projects. All are connected by the chain of responses to the water-transfer agreements between IID and some Southern California urban areas.

References

Asociación Ecológica de Usarios del Río Hardy-Colorado, A.C.; Center for Biological Diversity; Centro de Derecho Ambiental y Integración Económica de Sur, A.C.; Centro Intercultural de Estudios de Desiertos y Océanos, A.C.; Centro Regional de Estudios Ambientales y Científicos; Defenders of Wildlife; ECO-SOL Educación y Cultura Ecológica A.C.; Environmental Defense; El Grupo Ecológista Antares, A.C.; Friends of Arizona Rivers; High Country Citizens' Alliance; International Rivers Network; La Sociedad de Historia Natural Niparajá, A.C.; Living Rivers; Pacific Institute for Studies in Development,

Environment, and Security; Pro Esteros; ProNatura Peninsula de Baja California; Sierra Club; Colorado River Task Force; Sonoran Institute; and Southwest Rivers. 2001. "Binational Declaration: The Colorado River Delta." 12 September. http://www.biologicaldiversity.org/swcbd/Programs/watersheds/lcr/binational.pdf.

Blatter, J., and H. Ingram. 2000. "States and Markets Beyond: Governance of Transboundary Water Resources." *Natural Resources Journal* 40(2): 439–478.

Brown, C. P., and S. Mumme. 2000. "Applied and Theoretical Aspects of Binational Watershed Councils (Consejos de Cuencas) in the U.S.-Mexico Borderlands." *Natural Resources Journal* 40(4): 895–929.

Bustamante Redondo, J. 1999. *La Comisión Internacional de Límites y Aguas entre México y Los Estados Unidos*. Ciudad Juárez: Universidad Autónoma de Ciudad Juárez/San Diego State University.

Calleros, J. R. 1991. "The Impact on Mexico of Lining of the All-American Canal." *Natural Resources Journal* 31(4): 829–838.

Castro, J. L. 2002. "Sustainable Future of the Water Resources Along the U.S.-Mexico Border." Presentation to Canada's Public Policy Forum, Tijuana, 16 May. http://www.ppforum.ca.

Central Arizona Project. 1997. "Central Arizona Project." http://www.cap-az.com/about/index.cfm?action=cover.

Cohen, M. J., J. I. Morrison, and E. P. Glenn. 1999. *Haven or Hazard: The Ecology and Future of the Salton Sea*. Oakland, Calif.: Pacific Institute for Studies in Development, Environment, and Security. http://www.pacinst.org/reports/haven_or_hazard/index.htm.

Cortez Lara, A., and M. R. García Acevedo. 2000. "The Lining of the All-American Canal: The Forgotten Voices." *Natural Resources Journal* 40(2): 261–279.

Hayes, D. L. 1991. "The All-American Canal Lining Project: A Catalyst for Rational and Comprehensive Groundwater Management on the United States-Mexico Border." *Natural Resources Journal* 31(4): 803–827.

Mexican Senate. 2000. "Press Bulletin 2000/115." 27 April. http://www.senado.gob.mx.

Mendoza Luzcúber, O. J. 2002. "Un Problema Binacional Estados Unidos-México: Los Impactos del Cese de las Infiltraciones del Canal Todo Americano en la Agricultura de los Módulos de Riego 4, 5, 6 y 7 del Valle de Mexicali, B.C." Masters' thesis. El Colegio de la Frontera Norte/Centro de Investigación Científica y de Educación Superior de Ensenada, Tijuana, Baja California.

Michel, S. M. 2000. "Defining Hydrocommons Governance Along the Border of the Californias: A Case Study of Transbasin Diversions and Water Quality in the Tijuana-San Diego Metropolitan Region." *Natural Resources Journal* 40(4): 931–971.

Nitze, W. A. 2002. *Meeting the Water Needs of the Border Region. A Growing Challenge for the United States and Mexico.* Washington, D.C.: Center for Strategic and International Studies.

Ostrom, E. 2000. *El Gobierno de los Bienes Comunes.* Evolución de las Instituciones de Acción Colectiva. México, D.F.: Universidad Nacional Autónoma de México/Fondo de Cultura Económica.

Pensamiento, N. 2002. "Estados Unidos nos Daña con el Abasto Fronterizo del Agua." *El Heraldo de México* (16 April). http://www.rioweb.org/archive/year2002/april02/ehm_nosdanaabasto041602.html.

Pitt, J., D. F. Luecke, M. J. Cohen, E. P. Glenn, and C. Valdes-Casillas. 2000. "Two Nations, One River: Managing Ecosystem Conservation in the Colorado River Delta." *Natural Resources Journal* 40(4): 819–863.

San Diego County Water Authority. No Date. "San Diego County Water Authority." http://www.sdcwa.org.

Sánchez Rodríguez, R. 1990. *El Medio Ambiente como Fuente de Conflicto en la Relación Binacional México-Estados Unidos.* Tijuana: El Colegio de la Frontera Norte.

Utton, A. E. 1991. *The Transfer of Water from an International Border Region: A Tale of Six Cities and the All American Canal.* Albuquerque, N.M.: International Transboundary Resources Center, University of New Mexico.

Afterword

The old saying "the more things change, the more they stay the same" is aptly applied to water, especially along the U.S.-Mexican and California-Baja California borders. Over the decade-and-a-half of the Southwest Consortium for Environmental Research and Policy's (SCERP) investigation of border water issues, binational proposals by various agencies to address and solve challenges on both sides have come and gone. Other plans, problematic as they may be to binational relationships, persist and eventually reach fulfillment.

From the U.S. perspective, lining the All-American Canal seems an uncontroversial, if mildly, expensive way to deliver more water to thirsty coastal cities. Examined in a regional and long-term context, it is revealed to be contestable, contentious, and confounding. While a whole volume would be needed to properly explain the U.S. perspective, a few notes are warranted here to provide such context and to update issues raised in this volume.

As this Afterword is written in the last days of 2005, several lawsuits challenge the canal's lining. One pronounces that the Imperial Irrigation District (IID) did not have the right to sell farmers' water. After decade-long negotiations, IID had finally agreed to sell portions of its water rights to San Diego. The water would be "wheeled" through existing pipes from farther north on the Colorado River through Los Angles to San Diego. While the suit relates more to water saved by conservation and fallowing agricultural fields, it involves the water saved by lining the canal and contends that water made available through conservation should be applied to the most beneficial and equitable use and that transferring it out of the basin is neither. Another suit launched by California ecological groups contends that wetlands on the Mexican side of the border that harbor endangered species would dry up as a result of the lining.

Lining the All-American Canal: Competition or Cooperation for the Water in the U.S.-Mexican Border?

Local transborder cooperation on water issues waxes and wanes. The joint investigation of a binational aqueduct to convey water from the Mexicali-Imperial Valley to San Diego and Tijuana, as mentioned several times in the book, seemed a turn toward cooperation and even collaboration on regional water problems. But that potential joint effort was short lived. The San Diego County Water Authority (SDCWA) withdrew its support for such a project soon after a lengthy and elaborate feasibility study and immediately after a survey of users' perceptions of the binational project, which was never made public. Instead, the SDCWA is again pursuing unilateral desalination options at Encina and San Onofre, again after an inquiry into the relative costs and benefits of joint projects. While these projects may be significantly cheaper to build, they may portend to be much more expensive in the long-term due to energy costs and associated air quality impacts.

At the same time, SCERP sees encouraging signs of binational cooperation over water at the most local of scales. The California Resource Agency's Salton Sea recovery plan environmental impact review/statement carefully explains that all transboundary environmental impacts must be assessed and considered. Water districts immediately contiguous with the border are looking south to solve their water supply challenges. The entire Tijuana watershed is being considered the first transboundary basin to have a binational plan and a mechanism for addressing issues. A recently installed emergency connection allows either San Diego or Tijuana to get emergency supplies of water "wheeled" through the other in cases of disruption by earthquake or terrorism. Finally, plans to recover effluent from the International Wastewater Treatment Plant and convert it to useable water for Mexico are moving forward.

While the lining of the All-American Canal may move forward in early 2006, other water issues are being resolved with an eye across the border for cooperative opportunities to devise, design, and build better projects.

D. Rick Van Schoik
Managing Director, Southwest Consortium for Environmental Research and Policy

Index